海南热带雨林国家公园

刘俊 陈琛 著

华中科技大学出版社
http://press.hust.edu.cn
中国·武汉

内 容 简 介

你是否了解中国热带雨林的美丽？不妨跟随我们的脚步，从世界走向中国，探访我国最集中、类型最多样、保存最完好、连片面积最大的大陆性岛屿型热带雨林——海南热带雨林。这里建立了中国第一个"热带雨林"类型的国家公园。

海南热带雨林这个"宝库"生动展现了多种热带特有、中国特有、海南特有的自然奇观、明星动植物和民族文化。如此珍贵的"绿色宝藏"，随着国家公园的建设，得到了有效的修复、保护和可持续发展，逐步实现人与雨林和谐共生、共同发展的美好愿景。

让我们翻开读本，一起领略海南热带雨林国家公园的神奇魅力！

图书在版编目（CIP）数据

海南热带雨林国家公园 / 刘俊, 陈琛著. -- 武汉：华中科技大学出版社, 2024.6.

ISBN 978-7-5772-0957-9

Ⅰ. S759.992.66；S718.54

中国国家版本馆 CIP 数据核字第 20242QT562 号

海南热带雨林国家公园   刘俊 陈琛 著
Hainan Redai Yulin Guojia Gongyuan

项目总策划：李 欢
策 划 编 辑：胡弘扬 项 薇
责 任 编 辑：鲁梦璇
封 面 设 计：琥珀视觉
责 任 校 对：刘 竣
责 任 监 印：周治超
出 版 发 行：华中科技大学出版社（中国·武汉） 电话：（027）81321913
　　　　　　武汉市东湖新技术开发区华工科技园 邮编：430223
录　　　排：华中科技大学惠友文印中心
印　　　刷：湖北新华印务有限公司
开　　　本：787mm×1092mm　1/16
印　　　张：7.25
字　　　数：208 千字
版　　　次：2024 年 6 月第 1 版第 1 次印刷
定　　　价：98.00 元

本书若有印装质量问题，请向出版社营销中心调换
全国免费服务热线：400-6679-118　竭诚为您服务
版权所有　侵权必究

海南国家公园研究院科普系列丛书

序言一

  气候变化、生物多样性丧失、环境污染是当前人类社会面临的三大挑战。这些挑战不仅威胁着人类的生存环境，也给整个地球生态系统造成了巨大的威胁。

  为了应对这些挑战，国际社会已经采取了一系列的举措，包括制定《联合国气候变化框架公约》和《生物多样性公约》，旨在协调全球范围内的行动来减缓气候变化并保护生物多样性。在2022年的《生物多样性公约》第十五次缔约方大会上通过的"昆明－蒙特利尔全球生物多样性框架"中，190多个国家和地区承诺到2030年保护30%被认为对生物多样性至关重要的陆地和海洋，为全球生物多样性保护确立了更加明确的目标和路线图。

  中国作为全球最大的发展中国家，也积极行动了起来。在生态保护方面，中国正在从参与者、贡献者、追赶者变成引领者之一。习近平总书记提出了生态文明思想，并提出我国力争于2030年前实现"碳达峰"，努力争取在2060年前实现"碳中和"，中国的"双碳"目标是中国在应对全球气候变化方面对全世界作出的重大政策宣示。

  在生物多样性保护和国家公园建设方面，中国也取得了重要进展。中国设立了三江源、大熊猫、东北虎豹、海南热带雨林、武夷山首批5个国家公园，保护面积达23万平方千米，涵盖近30%的陆域国家重点保护野生动植物种类；同时，印发了《国家公园空间布局方案》，遴选出49个国家公园候选区，直接涉及省份28个，涉及现有自然保护地700多个，保护了超80%的国家重点保护野生动植物物种及其栖息地。

  2022年4月，习近平总书记来到海南热带雨林国家公园考察，强调海南要坚持生态立省不动摇，把生态文明建设作为重中之重，对热带雨林实行严格保护，实现生态保护、绿色发展、民生改善相统一；要跳出海南看这项工作，视之为"国之大者"，充分认识其对国家的战略意义，再接再厉把这项工作抓实抓好；强调海南热带雨林国家公园是国宝，是水库、粮库、钱库，更是碳库，要充分认识其对国家的战略意义，努力结出累累硕果。

海南虽然是较晚开展国家公园试点的省份，但是发展较快，海南热带雨林国家公园是习近平总书记亲自宣布的首批国家公园之一。海南国家公园研究院立足于为国家公园建设提供科技和智库支撑，目前已经积极努力做出了许多工作，取得了较显著的成绩。如针对濒危动物海南长臂猿保护的《海南长臂猿保护案例》在法国第七届世界自然保护大会上发布，并在第十五次生物多样性缔约方大会上做了"中国智慧、海南经验、霸王岭模式：海南长臂猿保护案例"专题报告，引起了国际社会的高度关注和赞扬，被国际社会誉为珍稀物种保护的中国智慧并建议向全球推广；又如，开展生物多样性保护和生物多样性关键区域（KBAs）研究，2022年与海南热带雨林国家公园管理局共同发布《海南热带雨林国家公园优先保护物种名录》，为全面开展热带雨林生物多样性保护和研究奠定了基础；再如，开展绿色名录等研究，为高质量建设海南热带雨林国家公园提供参考。

2018年4月13日，习近平总书记在庆祝海南建省办经济特区30周年大会上发表重要讲话，支持海南建设中国特色自由贸易港，着力打造全面深化改革开放试验区、国家生态文明试验区、国际旅游消费中心、国家重大战略服务保障区。正是在这样的背景下，海南国家公园研究院编写了以国家公园和生物多样性保护为主题的系列科普丛书。我们希望通过这套丛书，向公众普及环保知识，激发人们对保护自然的关注和参与。同时，我们也希望借助这些书籍，让更多人了解中国在生态文明建设方面的努力和成就，为构建人与自然和谐共生的美好未来贡献力量。

海南国家公园研究院在国家公园建设中不仅提供科技支撑，同时作为智库，致力于普及科学知识与提升公众的科学素养。希望通过阅读这一系列的科普读物，大家能有所收获并且能积极行动起来，为国家公园建设和生物多样性保护事业贡献一份力量，让人与自然和谐共生的美好画卷率先在海南实现，让我们的地球变得更加美丽，让生态保护的成果惠及全人类！

世界自然保护联盟原总裁兼理事会主席
174—177届联合国教科文组织执行理事会主席　　**章新胜**
海南国家公园研究院资深专家

海南国家公园研究院科普系列丛书

序言二

  科学普及是一座桥梁，连接着知识的源泉与广大民众。习近平总书记高度重视科学普及工作，多次强调"科学普及是实现创新发展的重要基础性工作""科技创新、科学普及是实现创新发展的两翼，要把科学普及放在与科技创新同等重要的位置"。党的二十大将科普教育作为生态文明建设的一个重要组成部分；《关于新时代进一步加强科学技术普及工作的意见》明确提出"推动科普全面融入经济、政治、文化、社会、生态文明建设"。这明确了科普工作的历史使命和时代要求，为新时代科普高质量发展指明了方向。

  作为科学普及的重要载体和路径，科普教育的重要性不言而喻。它承载着知识传播的作用，更是文明程度提升、科技发展的基石。此外，科普教育对国民素质的提升至关重要。它不仅让人们了解科学知识，更培养了批判性思维、创新能力和科学探索精神。随着生态文明建设的大力推进，国家公园的科普教育工作受到越来越多的关注。

  海南热带雨林国家公园内分布有我国最集中、类型最多样、连片面积最大、保存最完好的大陆性热带雨林，蕴含着丰富的生态系统和生物多样性资源，是一个天然的博物馆，也是一座宝贵的天然科普教育基地。它不仅具有重要的休憩和体验价值，更是一个蕴含了丰富知识和信息的宝藏。在这里，我们可以近距离感受到热带雨林的神秘和壮美。

  海南热带雨林国家公园是生物多样性和遗传资源的宝库。已有的调查和研究显示，截至2023年，公园内分布有野生维管束植物4367种、野生脊椎动物651种，其中特有维管束植物419种、特有陆生脊椎动物23种。公园内分布有全球仅存的42只海南长臂猿，具有极高的保护价值，是海南乃至中国生物多样性保护的靓丽名片。海南热带雨林国家公园也是海南岛的生态安全屏障，具有重要的水源涵养、固碳释氧、土壤保持、气候调节和防灾减灾等功能。习近平总书记视察海南时强调："热带雨林国家公园是国宝，是水库、粮库、钱库，更是碳库。"

海南国家公园研究院是国家公园的智库支撑，长期致力于研究这一"国之大者"，运用自身的知识和技术优势，积极推动科普教育。这套丛书通过通俗易懂的语言，将海南热带雨林丰富的自然景观和生态价值讲述给更多的人，让更多人领略到热带雨林生态系统的壮美，提升公众的民族自豪感和国家认同感，让公众与这片具有国家代表性的自然保护地建立紧密的联系。

让我们一起探索海南热带雨林的奥秘，共同了解、关注、参与保护这片自然奇迹！让科学知识走进千家万户，让每个人都成为科学的传播者和科普的受益者。

中国工程院院士　**杨志峰**

海南国家公园研究院科普系列丛书

前言

亲爱的读者朋友：

您好！

国家公园是我国最重要的自然生态空间，是自然生态系统中最重要、自然景观最独特、自然遗产最精华、生物多样性最富集的部分，极具保护价值。我国实行国家公园体制，是推进自然生态保护、建设美丽中国、促进人与自然和谐共生的一项重要举措。

位于海南岛中部的海南热带雨林国家公园，是我国首批设立的五个国家公园之一，这里拥有我国分布最集中、类型最多样、保存最完好、连片面积最大的大陆性岛屿型热带雨林，拥有海南、中国乃至世界独有的动植物及种质基因库，是"水库""粮库""钱库"，更是"碳库"，也是多种珍稀濒危动植物的庇护所，被世界自然保护联盟（IUCN）物种红色名录列为极危（CR）的国家一级保护野生动物海南长臂猿就生活在这里。可以说，海南热带雨林国家公园是一个天然的"博物馆"，是国宝。

为了让大家更好地认识海南热带雨林国家公园这个国宝，提升大家的生态环保意识和科学素养，进而让大家参与到保护海南热带雨林的行动当中，最终实现生态保护、绿色发展和民生改善相统一，在中共海南省委宣传部、海南省旅游和文化广电体育厅、海南省科学技术协会的亲切关怀和悉心指导下，海南省林业局（海南热带雨林国家公园管理局）、海南省教育厅、共青团海南省委、海南省文学艺术界联合会、联合国教科文组织（UNESCO）驻华代表处和海南国家公园研究院共同主办为期五年的"助力双碳目标，保护热带雨林"科普教育系列活动。本次我们委托华南师范大学和海南师范大学专业团队编写的科普读物《海南热带雨林国家公园》和《元元说猿：漫话海南长臂猿》，便是该系列活动的一部分。

本套丛书的定位是轻科普，不仅注重知识的科学性、准确性，更注重趣味性和可读性，旨在向全年龄段人群普及国家公园和生物多样性保护的知识。无论您是小学生、大学生、职场人士，还是退休的老年朋友，这套丛书都将为您提供有趣、易懂的科学知识。

海南热带雨林国家公园的珍稀物种、自然景观和人文历史相辅相成，共同构成了一幅美丽的画卷，我们深知科学知识的普及对于推动生物多样性和国家公园保护、提升大众的自然和人文素养具有重要作用。因此，我们精心设计了这套书籍，力求用简洁易懂的语言和生动有趣的插图，深入浅出地为读者讲解海南长臂猿等生物多样性保护和国家公园建设的重要价值和意义。

之后，我们将逐步推出更多专业性与趣味性相融合的书籍。无论是对环境保护事业感兴趣的普通读者，还是对生态学、环境科学等专业领域有所涉猎的专家学者，我们都将力求提供丰富多样的内容，满足不同读者的需求。

海南热带雨林国家公园，作为一个天然博物馆，将成为我们探索的起点。这里不仅具有很高的观赏价值，更是一个充满学习和探索机会的地方。我们将带您一起走进这片国宝，探寻其中的奥秘。

在本套丛书的编写过程中，我们得到了海南热带雨林国家公园管理局各分局、五指山市人民政府，以及琼中、白沙、东方、陵水、昌江、乐东、保亭、万宁各市县提供的帮助；得到了国家社会科学基金项目《知识转移促进自然保护地生态产品价值实现的机制研究》（2023—2026）（批准号：23BJY141）给予的理论和智力支撑；得到了联合国教科文组织东亚地区办事处、中国青少年发展基金会梅赛德斯-奔驰星愿基金以及海南绿岛热带雨林公益基金会的支持。在此谨向各支持单位和个人表示衷心感谢。

我们希望通过本套系列丛书的推出，唤起更多人对海南长臂猿等生物多样性保护和国家公园建设的关注，推动更多人士参与到人与自然和谐共生的行动中来。由于编写水平和时间有限，书中难免有不足和疏漏之处，也欢迎广大读者对这套丛书提出宝贵建议，使我们能够不断提高编写水平。

让我们携手努力，共同创造一个更加美好的地球家园！

海南国家公园研究院
执行院长 教授　**汤炎非**

目录

第一章　热带雨林　地球宝库　　2
Chapter 1 Tropical Rainforest: Earth's Treasure Trove
　　什么是热带雨林？　　3
　　What is the Tropical Rainforest?
　　中国热带雨林在哪里？　　6
　　Where are the Tropical Rainforests in China?
　　"四库"国宝：海南热带雨林　　10
　　The Four Great Treasuries of Hainan Tropical Rainforest

第二章　中国国家公园来了！　　20
Chapter 2　China's National Parks are Coming！
　　国家公园诞生记　　21
　　The Birth of National Parks
　　世界国家公园　　24
　　World's National Parks
　　中国国家公园　　30
　　China's National Parks

第三章　探秘海南热带雨林国家公园　　38
Chapter 3　Exploring the National Park of Hainan Tropical Rainforest

海南热带雨林国家公园的显著特点　　40
The Distinguished Features of the National Park of Hainan Tropical Rainforest

热带雨林秘境有奇观　　48
Abundance of Wonders in the Tropical Rainforest Secrets

猿声深处万物生　　58
Untouched Wilderness Bawangling

峻岭飞瀑山水秀　　66
Steep Mountains and Splendid Waterfalls

黎苗人家风情浓　　74
Rich Cultural Charm of Li and Miao Ethnic Groups

第四章　爱在雨林　　80
Chapter 4 Love In the Rainforest

雨林生态正在修复　　82
Rainforest Ecosystem on the Path to Recovery

雨林与您　　88
Rainforest and You

开启探索之旅

第一章
热带雨林　地球宝库

Chapter 1
Tropical Rainforest: Earth's Treasure Trove

什么是热带雨林？
What is the Tropical Rainforest?

五指山 图摄 / 孟志军

地球腰带上的森林
Forests on Earth's Waistband

热带雨林大多位于赤道多雨气候区（赤道到南北纬10°之间）、热带季风气候区（北纬10°到北回归线附近），这里长年气候炎热、雨水充足，生物在这里快速生长、种类繁密，热带雨林像一条绿色腰带挂在地球上。

The majority of tropical rainforests are located in regions with wet equatorial climates (between the equator and 10° latitude) and tropical monsoon climates (from around 10° N latitude to near the Tropic of Cancer). The climate in these areas is hot throughout the year with abundant rainfall, fostering rapid and dense growth of diverse species. It's like a green belt hanging around the Earth.

The Three Major Tropical Rainforests in the World

In this belt, there are three major tropical rainforests: the Amazon Rainforest, the Congo Rainforest and Basin, and the Asian Rainforest.

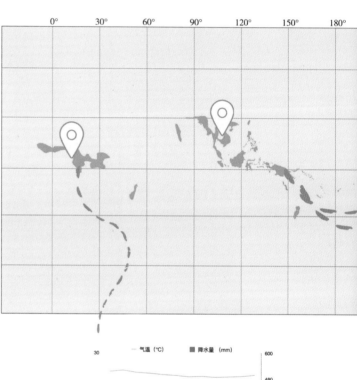

World's Tropical Rainforest Distribution

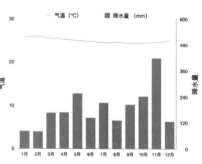

刚果盆地热带雨林平均气温降水图（20°E，0°）（2020—2022年）

The Congo Rainforest and Basin Average Temperature and Precipitation Chart (20°E, 0°) (2020—2022)

刚果盆地热带雨林是非洲森林之王，这里也是人猿泰山和他的兄弟大猩猩的家。

The Congo Rainforest and Basin is the king of African forests. It's also the home of Tarzan and his brothers, the gorillas.

世界三大热带雨林

在这条腰带上，散落着三大热带雨林：亚马孙热带雨林、刚果盆地热带雨林、亚洲热带雨林。

世界热带雨林分布

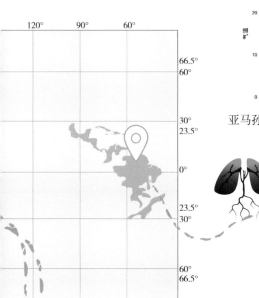

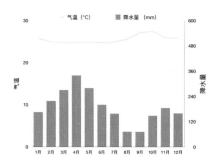

亚马孙热带雨林平均气温降水图（56°W，0°）
（2020—2022年）
The Amazon Rainforest Average Temperature and Precipitation Chart (56°W, 0°) (2020—2022)

亚马孙热带雨林是"老大"，面积和"肺活量"都最大，被称为"地球之肺"。

The Amazon Rainforest is "the big brother". It has the largest area and "vital capacity", and is known as the "Lungs of the Earth".

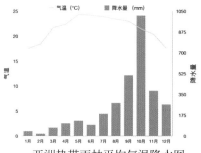

亚洲热带雨林平均气温降水图
（108°E，15°N）（2020—2022年）
The Asian Rainforest Average Temperature and Precipitation Chart
(108°E, 15°N) (2020—2022)

亚洲热带雨林是"小弟"，这里珊瑚礁、岛屿、海滩特别多，最受世界各地的游客喜爱。

The Asian Rainforest is "the little brother". There are a lot of coral reefs, islands, and beaches, making it a favorite among tourists worldwide.

"长翅膀"的龙脑香科

"Flying" Dipterocarpaceae

热带雨林是全球最大的生物基因库，生长着数以万计的生物。其中，龙脑香科植物是亚洲热带雨林的标志性物种。中国正是因为找到了龙脑香科植物，才确定本土也有热带雨林。你能想到吗？这么一棵高50米的大树上长着一堆长翅膀的小果子，风一吹，这些果子好像小叮当的竹蜻蜓，能带你去任何地方。

The tropical rainforest is the world's largest biological gene bank, home to tens of thousands of species, as well as microorganisms. Among them, the Dipterocarpaceae is the iconic species of the Asian Rainforest. China was able to confirm the existence of tropical rainforests within its borders because it discovered the Dipterocarpaceae family of plants. Can you believe it? On these enormous 50-meter-tall trees, you'll find a bunch of little fruits with long wings. When the wind blows, they look like Doraemon's Take-copter and can take you anywhere.

中国热带雨林在哪里?

Where are the Tropical Rainforests in China?

热带雨林内部 图摄 / 卢刚

中国的热带雨林主要分布于台湾南部、海南、广西、广东和云南南部及西藏东南的部分地区。这些地方处于亚洲热带地区的边缘，主要为热带季风气候条件下发育的雨林，区别于赤道雨林。中国的热带雨林分布较少，它们的面积仅占国土总面积的 3.2%，却拥有中国物种总数的 25%，生态系统类型的 25.8%，是名副其实的生物基因库啊！

China's tropical rainforests are mainly found in southern Taiwan, Hainan, Guangxi, Guangdong and southern Yunnan and parts of southeastern Tibet. These areas are on the edge of tropical Asia, which is mainly developed under the tropical monsoon climate, different from the equatorial rainforest. Tropical rainforests in China are limited in distribution. These areas only cover 3.2% of the total land area in China, but they possess 25% of the total number of species and 25.8% of ecosystem types in the country. They are truly a biological gene bank!

中国热带雨林

China's Tropical Rainforest

让我们把目光聚焦到海南热带雨林！

Let's turn our attention to the Hainan tropical rainforest!

虽然海南热带雨林是我国离赤道最近的热带雨林，但它位于世界雨林分布的北缘(108°44'32"—110°04'43" E，18°33'16"—19°14'16" N)，因此其植被富于热带性，但有别于赤道植被，具有热带季风气候植被的特点。

Although the Hainan tropical rainforest is the closest tropical rainforest to the equator in our country （108°44' 32" —110°04' 43" E, 18°33' 16"—19°14' 16" N), it is also at the northern edge of the world's rainforest distribution, so its vegetation is rich in the tropics but distinct from equatorial vegetation, characterized by monsoon tropical vegetation.

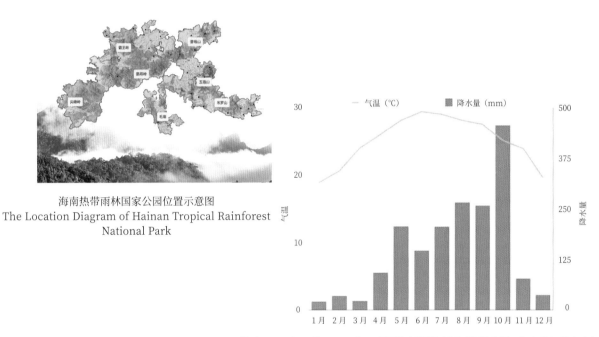

海南热带雨林国家公园位置示意图
The Location Diagram of Hainan Tropical Rainforest National Park

海南 2020 年 6 月—2023 年 5 月平均气温降水图（数据来源：海南省气候中心）
Average Temperature and Precipitation Chart for Hainan from June 2020 to May 2023 (Data Source: Hainan Provincial Climate Center)

海南这片热带雨林的自然生态系统具有独特性和原真性，所以它被设立为我国第一个热带雨林类型的国家公园。

The natural ecosystem of the Hainan tropical rainforest is unique and authentic, making it the first national park of tropical rainforest type in China.

这里许多动植物资源都被贴上了"中国特有、热带特有、世界特有"的标签！

Many of the flora and fauna here have been labeled as "endemic to China, the tropics and the world"!

"四库"国宝：海南热带雨林
The Four Great Treasuries of Hainan Tropical Rainforest

"Tropical Rainforest National Parks are national treasures, serving as reservoirs, grain depots, vaults, and also carbon pools. We should fully recognize their strategic significance to our country and strive to yield abundant results."　　Xi Jinping

水库　　Reservoirs

你可以把雨林看成自然界的绿色大水库，这个水库就像是……
You can think of rainforests as a natural green reservoir, and this reservoir is just like...

雨水储存缸
Rain water Storage Tank

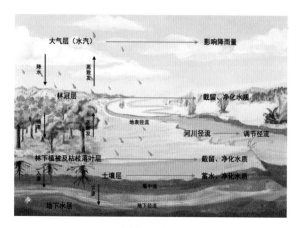

水循环示意图

老天爷每天向地面打开水龙头，雨林照单全收。通过雨林中的河流、湖泊、生物、土壤等，把水分储存起来。雨林涵养水源的能力超乎你的想象。亚马孙雨林涵养的水量约占地球表面淡水总量的 1/4，流域内近 1/2 的降水也是由雨林自己产生的。

God opens the tap to the ground every day, and the rainforest lapsit up. Water is stored through rivers, lakes, organisms, soil, etc., in the rainforest. The rainforest's capacity to conserve water is beyond your imagination. The Amazon Rainforest conserves the amount of water that accounts for about 1/4 of the total amount of freshwater on the surface of the Earth, nearly 1/2 of the precipitation in the watershed is also produced by the rainforest itself.

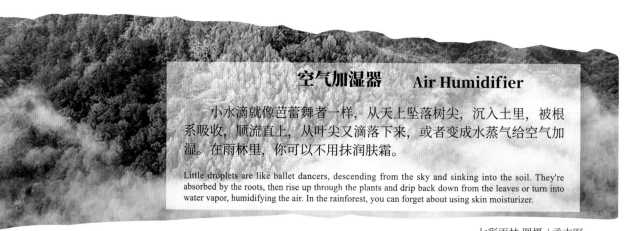

空气加湿器　　Air Humidifier

小水滴就像芭蕾舞者一样，从天上坠落树尖，沉入土里，被根系吸收，顺流直上，从叶尖又滴落下来，或者变成水蒸气给空气加湿。在雨林里，你可以不用抹润肤霜。

Little droplets are like ballet dancers, descending from the sky and sinking into the soil. They're absorbed by the roots, then rise up through the plants and drip back down from the leaves or turn into water vapor, humidifying the air. In the rainforest, you can forget about using skin moisturizer.

七彩雨林 图摄 / 孟志军

十里画廊 图摄 / 孟志军

水量调节闸
Water Flow Regulating Gate

当雨水丰富的时候，雨林就发挥"储存缸"的作用，把雨水存进地下和植被中。当旱季来临，雨林就"开闸放水"，给河流补充水分。既降低了洪涝风险，又确保了野生动物和人有足够的水资源。

When rainfall is abundant, the rainforest acts like a "storage tank", storing rainwater underground and within its vegetation. When the dry season arrives, the rainforest "opens the floodgates", replenishing rivers with water. This not only reduces the risk of flooding but also ensures that both wildlife and people have an ample water supply.

水质净化器
Water Quality Purifier

落下来的雨水，被雨林高大的乔木一层一层拦截，密密麻麻的植物根系不断过滤水中杂质，就像家里的净水器，对改善水质起到了至关重要的作用。

The falling rain is intercepted by the towering trees of the rainforest, and the intricate web of plant roots continually filters out impurities in the water, just like the filter in your home water purifier, playing a crucial role in improving water quality.

水质净化示意图

粮库

热带雨林是陆地上生物多样性水平最高的区域,这里生存着 80% 以上的生物物种,1/4 的现代药物来源于热带雨林植物。

Tropical rainforests are the regions on land with the highest levels of biodiversity, hosting over 80% of the world's biological species. 1/4 of modern medicines are derived from plants in tropical rainforests.

海南石斛 图摄 / 孟志军

亚马孙热带雨林

亚马孙热带雨林是目前全球面积最大的原始森林。其总面积为 700 万 km^2,占全球热带雨林的 1/3。雨林内有 1000 万种生物,生物约占全球生物总数的 1/3。植物约有 500 万种,约占全球植物总数的 10%。穿过雨林的全球第一大河亚马孙河内有 3000 余种鱼类,数量是整个欧洲大陆的 15 倍。亚马孙热带雨林是当之无愧全球最大的基因宝库、粮食宝库。

The Amazon Rainforest is currently the world's largest primary forest, covering a total area of 7 million square kilometers, which accounts for one-third of the world's tropical rainforests. Within the rainforest, there are approximately 10 million species, representing about one-third of the world's total biodiversity. It is home to around 5 million plant species, constituting approximately 10% of the world's total plant diversity. The Amazon River, the world's largest river that flows through the rainforest, harbors over 3000 species of fish, which is 15 times more than the entire European continent. The Amazon Rainforest is rightfully considered the world's largest genetic and food resource repository.

Grain Depots

亚马孙热带雨林每公顷的生物种类
Biological Species per Hectare of the Amazon Rainforest

种类 Type	树木 Trees	鱼类 Fishery	鸟类 Birds	哺乳动物 Mammal
数量 Quantity	300 种	1400 余种	1300 种 （191种是世界独有的）	300 种

数据来源：莫鸿钧《巴西亚马孙河流域生物多样性和生态环境的保护对策》，《中国农业资源与区划》，2004 年第 3 期。

海南热带雨林是全球重要的种质资源基因库
The Hainan Tropical Rainforest is a Globally Significant Genetic Resources Gene Bank

类别 Category	种类 Type	占比 % Proportion		国家重点保护种类 National Key Protected Species	国家一级 First-Class	国家二级 Second-Class
		全省 Province	全国 Country			
野生维管束植物 210 科 1159 属 Wild vascular plants, 210 families, 1159 genera	3653	77.91	11.7	149	6	143
陆栖脊椎动物资源 5 纲 38 目 145 科 414 属 Terrestrial vertebrate resources, 5 classes, 38 orders, 145 families, and 414 genera	540	77.36	18.62	145	14	131

数据来源：《海南热带雨林国家公园总体规划（2022—2030 年）》

海南热带雨林 图摄 / 孟志军

钱库

绿水青山怎么转换成"金山银山"？
How Can Lucid Waters and Lush Mountains be Transformed into Invaluable Assets?

我们用国内生产总值（Gross Domestic Product，简称 GDP）来衡量一个地区的经济发展实力。
我们用生态产品总值（Gross Ecosystem Product，简称 GEP）来衡量一个地区的生态产品价值。
GEP 是指一定区域内各类生态系统在核算期内提供的所有生态产品的货币价值之和。这一指标能更清晰地显现一个区域的"绿色家底"，为我们将绿水青山转换成"金山银山"、保护和建设"绿色银行"提供帮助。

We measure the economic product value of a region using Gross Domestic Product (GDP).
To gauge the level of ecological civilization in a region, we use Gross Ecosystem Product (GEP).
GEP refers to the total monetary value of all ecological products provided by various ecosystems within a specific area during the accounting period. This indicator can more clearly reveal the green foundation of a region, helping us transform lucid waters and lush mountains into invaluable assets and assisting in the production and construction of the "green bank".

物质产品：农业产品、林业产品、畜牧业产品、渔业产品、生态能源和其他。

Material Products: agricultural products, forestry products, husbandry products, fishery products, eco-energy and others.

海南山灵芝 图摄 / 曾念开

白查村 图摄 / 孟志军

玉蕊 图摄 / 孟志军

用数据衡量生态环境不是一件容易的事，海南热带雨林国家公园在全国率先开展国家公园范围内的 GEP 核算，并成为首个发布 GEP 核算成果的国家公园。

Accounting for the ecological environment is not an easy task. Hainan Tropical Rainforest National Park has taken the lead in the country in carrying out GEP accounting within the scope of national parks, and has become the first national park to release the results of GEP accounting.

2021 年，海南省发布 2019 年度海南热带雨林国家公园的生态系统生产总值（GEP）为 2045.13 亿元，这片生态宝地首次拥有了自己实实在在的"身价"。

2023 年，海南省海南热带雨林国家公园 2021 年度生态系统生产总值（GEP）结果新鲜出炉，为 2068.39 亿元。其中，生态系统物质产品价值（包含农业产品、林业产品、畜牧业产品、生态能源等 6 项指标）为 49.96 亿元，占国家公园 GEP 总量的 2.42%；生态系统调节服务价值（包含涵养水源、生物多样性、固碳释氧、洪水调蓄和空气净化等 9 项指标）为 1728.81 亿元，占 83.58%；生态系统文化服务价值（包含休闲旅游、景观价值等 4 项指标）为 289.62 亿元，占 14%。

Vaults

生态产品价值（GEP）怎么算？
How to Calculate GEP?

五指山 图摄 / 孟志军

黎母山 图摄 / 孟志军

海南黎族 图摄 / 孟志军

大型真菌调查 海南省林业局供图

吊罗山 图摄 / 孟志军

调节服务：涵养水源、保育土壤、固碳释氧、空气净化、森林防护、洪水调蓄、气候调节、生物多样性和授粉服务。

Regulating Services: Water source conservation, soil conservation, carbon sequestration and oxygen release, air purification, forest protection, flood regulation and storage, climate regulation, biodiversity and pollination services.

文化服务：休闲旅游、景观价值、科学研究和科普教育。公园内黎族传统文化是区别于其他地区、其他国家公园的特有人文资源，向人们展示了其独特气质和底蕴。

Cultural Services: Leisure tourism, landscape value, scientific research and popular science education. The traditional culture of the Li ethnic groups in the park is a unique human resource that distinguishes it from other regions and other national parks, showing people its unique temperament and heritage.

In 2021, Hainan Province released that the 2019 Hainan Tropical Rainforest National Park's Gross Ecosystem Product (GEP) was 204.513 billion yuan. This ecological treasure land has its own real "value" for the first time.

In 2023, the 2021 Gross Ecosystem Product (GEP) result for Hainan Tropical Rainforest National Park in Hainan Province was freshly released at 206.839 billion yuan. Among these, the value of ecosystem material products (including six indicators such as agricultural products, forestry products, husbandry products, and ecological energy) was 4.996 billion yuan, accounting for 2.42% of the total GEP of the national park. The value of ecosystem regulation services (including nine indicators such as water source conservation, biodiversity, carbon sequestration and oxygen release, flood regulation and storage, and air purification) was 172.881 billion yuan, accounting for 83.58%. The value of ecosystem cultural services (including four indicators such as leisure tourism and landscape value) was 28.962 billion yuan, accounting for 14%.

碳库

截至 2020 年，全球森林碳储量高达 6620 亿吨，约占全球植被碳储量的 77%。热带雨林植被密度高、种类丰富，起到了强大的"碳库"作用，也展现了强大的碳汇能力！

The global forest carbon storage is as high as 662 billion tons, accounting for about 77% of the global vegetation carbon storage. Tropical rainforests have high vegetation density and rich species, playing a strong role as a "carbon pool" and demonstrating strong carbon sink capacity!

雨林是如何固碳的？
How do Rainforests Sequestr Carbon?

热带雨林植被 图摄 / 孟志军

简单来说，雨林里的碳汇是指雨林茂密的植被、宽大的叶子通过光合作用，将人类生产、生活和动物排泄、微生物分解作用等释放到空气中的二氧化碳吸收转化为碳水化合物，最后储存在树干、树枝、树叶和根部等各个角落。

In simple terms, the carbon sinks in the rainforest refer to the dense vegetation and large leaves of rainforests absorb carbon dioxide released into the air through human production, life, animal excretion, microbial decomposition, etc., through photosynthesis, convert it into carbohydrates, and store it in various parts of the trunk, branches, leaves, and roots.

海南尖峰岭森林生态系统国家野外科学观测研究站（以下简称尖峰岭生态站）站长陈德祥等人曾做过一项研究，计算出尖峰岭热带雨林取样树种各器官的生物量含碳率——树干 51.08%、树根 50.54%、树叶 50.38%、树枝 49.40%。所以保护雨林，就是增加碳汇"仓储"。

Chen Dexiang, the director of Hainan Jianfengling Forest Ecosystem National Field Science Observation and Research Station (hereinafter referred to as Jianfengling Ecological Station), and others have conducted an experiment to calculate the carbon content rate of each organ of the sample trees in Jianfengling Tropical Rainforest—trunk 51.08%, root 50.54%, leaf 50.38%, branch 49.40%. Therefore, protecting rainforests is to increase the storage of carbon sinks.

碳循环过程示意图

Carbon Pools

雨林在碳汇交易中有什么价值?
What is the Value of Rainforests in Carbon Sink Trading?

碳汇交易是基于《联合国气候变化框架公约》及《京都议定书》对各国分配二氧化碳排放指标的规定，创设出来的一种虚拟交易。简单来说，就是碳排放量高的发达国家可以向发展中国家购买碳排放指标，或者在发展中国家投资造林以增加碳汇、抵消碳排放的交易机制。

Carbon sink trading is a virtual transaction created based on the provisions of the *United Nations Framework Convention on Climate Change* and the *Kyoto Protocol* for the allocation of carbon dioxide emission targets to countries. Simply put, developed countries with high carbon emissions can purchase carbon emission targets from developing countries or invest in afforestation in developing countries to increase carbon sinks and offset carbon emissions through this trading mechanism.

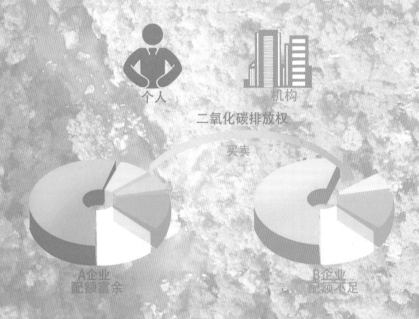

碳汇交易示意图

林业碳汇就是一个"可量化、可交易、可增值、可持续"的商品，当它进入碳交易市场时，会产生额外的经济价值，成为"绿水青山变为金山银山"的重要途径，也推动着保护者受益、使用者付费、破坏者赔偿的利益导向机制逐步形成。

The forestry carbon sink is a "quantifiable, tradable, value-added and sustainable" commodity, when it enters the carbon trading market, it will generate additional economic value and become an important way to "Lucid waters and lush mountains transform into invaluable assets". It also promotes the gradual formation of the interest-oriented mechanism of the protector's benefit, the user's pay, and the destroyer's compensation.

雨林俯瞰 吊罗山分局供图

第二章
中国国家公园来了！

Chapter 2
China's National Parks are Coming！

国家公园诞生记
The Birth of National Parks

19世纪30年代，美国人大搞西进运动，拓荒掘金，在地里挖出了石油、矿石等宝贵资源。乘着第一次工业革命后期的"西风"，美国大力发展工业，一跃成为世界经济大国，但同时也破坏了大量自然资源与生态环境。

In the 1830s, Americans embarked on the Westward Expansion. They ventured westward, engaging in activities like pioneering and gold mining. In the process, they discovered valuable resources such as oil and minerals, and harnessed the industrial power of the late stages of the Industrial Revolution. This rapid industrialization transformed the United States into an economic superpower. However, it also exacted a toll on natural resources and the environment, causing significant damage.

与此同时，一位美国东部的著名画家——乔治·卡特林，热衷于印第安人物和文化的绘画题材，关注印第安文化保存。在采风的过程中，他看到了西进运动对印第安部落生存环境和文化的毁灭性影响。为了保护美洲原住民文化和他们的生活环境，身为律师的乔治·卡特林在1832年提出了"国家公园"的概念，并用法律途径直接向美国国会建言建立国家公园，这就是国家公园理念的起源。后人把他视为国家公园理念的提出者。

Simultaneously, a renowned Eastern American artist named George Catlin, known for his passion for painting subjects related to Native Americans and their culture, developed a concern for the preservation of Native American culture. While traveling to paint these subjects, he witnessed the devastating impact of Westward Expansion on the survival environment and culture of Native American tribes. In 1832, George Catlin, also a lawyer, introduced the concept of "national parks" to protect the culture and living environment of the Native Americans. He took a legal approach to directly advise the United States Congress to establish national parks, which marked the inception of the national park concept. He is regarded by later generations as the presenter of the idea of national parks.

　　在如此优美壮阔的园地中，保留着自然之美与原始风貌，在这里，世人看到印第安人一路走来的足迹。在这样一个国家公园里，人与万物共存，充满着最原始的惊艳。——乔治·卡特林

In such a magnificent and expansive domain, the beauty and primordial essence of nature are preserved. It is a place where people can witness the footsteps of Native Americans throughout their journey. In such a national park, the coexistence of humans and all living beings is harmonious, brimming with the most primitive and astonishing wonders. —George Catlin

　　直到40年后的1872年，美国黄石国家公园正式建立，才让"国家公园"的理念得以实现。

It wasn't until 1872, 40 years later, that the establishment of Yellowstone National Park in the United States finally brought the concept of "national parks" to fruition.

黄石国家公园标志牌

23

世界国家公园

World's National Parks

美国：黄石国家公园
United States: Yellowstone National Park

1872 年，世界上第一个国家公园——黄石国家公园诞生了。

In 1872, the world's first national park, Yellowstone National Park, was born.

它坐落在"美洲的脊梁"落基山脉。这里就像大自然的宝藏盒，你可以观察地球的"心跳"，看世界上最壮观的破火山口，数一数上百个间歇泉喷出热泉的节奏，还可以来一场发现棕熊、灰狼的探险之旅。

It's nestled in the "backbone of the Americas", the Rocky Mountains. Here is like Mother Nature's treasure chest. You can observe the Earth's "heartbeat", witness the most spectacular caldera in the world, count the rhythms of over a hundred geysers spouting hot springs, and embark on an adventure to discover brown bears and gray wolves.

火棱镜彩泉
The Grand Prismatic Spring

下瀑布
Lower Falls

老忠实间歇泉
Old Faithful

肯尼亚：马赛马拉国家公园

Kenya: Masai Mara National Reserve

"哈库那，玛塔塔"（没有烦恼，无忧无虑）。
你还记得《狮子王》里的那句快乐咒语吗？
欢迎来到地球上最有野性的国家公园——肯尼亚马赛马拉国家公园。

"Hakuna Matata" (which means "No worries" in Swahili). Do you remember that happy spell from "The Lion King"? Welcome to Masai Mara National Reserve, one of the wildest national parks on Earth!

每年的7—8月，马赛马拉的草原都会上演"天国之渡"，这是世界上最狂野的食草动物大迁徙。近250万头食草动物，为了鲜嫩的食物，每年沿着东非大草原，按顺时针方向迁徙。每年的6—9月，马赛马拉大草原上的青草最为鲜嫩，但这里的面积只占整个东非大草原的1/11，因此在这短短的4个月内，250万头食草动物会分批聚集在此，想象一下这个狂野而震撼的场面！

Every year in July and August, the grasslands of Masai Mara witness the Great Migration, which is the wildest herbivore migration in the world. Almost 2.5 million herbivores migrate clockwise along the East African savannah in search of fresh and tender vegetation. From June to September, Masai Mara offers the greenest and freshest grass, but this area covers only 1/11th of the entire East African savannah. Therefore, within these four months, the 2.5 million herbivores gather here in batches. Imagine this awe-inspiring and wild spectacle!

荷兰：高费吕沃国家公园

The Netherlands: De Hoge Veluwe National Park

　　如果说世界上收藏梵高作品最多的地方是位于荷兰阿姆斯特丹的梵高博物馆，那么第二多的就是高费吕沃国家公园了，这里又名"梵高国家森林公园"。

If the Van Gogh Museum in Amsterdam, the Netherlands, has the most extensive collection of Van Gogh's works in the world, the second-largest collection is at the De Hoge Veluwe National Park, also known as the "Van Gogh National Forest Park".

　　这里有深邃幽静的森林，也有长满石南的原野，甚至还有流动的沙丘，是荷兰保护最好的国家公园，记录了珍贵的地质演化历史，成为荷兰非常受欢迎的旅游景点之一。这里的库勒慕勒博物馆也是世界上梵高作品第二大展示地。逛逛森林，看看名作，十分惬意！

Here, you can find deep and tranquil forests, fields covered in heather, and even shifting dunes. It's the best-preserved national park in the Netherlands, recording valuable geological history and becoming one of the country's most popular tourist destinations. The Kröller-Müller Museum here is the second-largest display of Van Gogh's works in the world. Take a stroll through the forest, enjoy the famous artworks, and have a splendid time!

世界上第一个国家公园——黄石国家公园的设立，带动了世界范围的国家公园建设，许多国家纷纷效仿黄石国家公园的模式，建立属于自己的国家公园。根据世界保护区数据库（WDPA）2023年1月公布的自然保护地数据，目前全球国家公园约有6000处。

The establishment of the world's first national park, Yellowstone National Park, drove the construction of worldwide national parks, with many countries following the model of Yellowstone National Park to establish their own national parks. According to the World Database on Protected Areas (WDPA) released in January 2023, there are currently about 6000 national parks around the world.

从北美的荒野保护到欧洲的自然游憩，国家公园的建立，不仅是为了保护自然美景、多元的生态环境、独特的地质遗产，还承担着保护民族历史文化、开展教育、发展旅游等责任。

From wilderness protection in North America to nature-based tourism and recreation in Europe, national parks have been established not only to protect natural beauty, diverse ecosystems, and unique geological heritage but also to take on responsibilities such as preserving national history and culture, carrying out education, and developing tourism.

现阶段，我国国家公园建设还处于起步阶段，研究和借鉴世界上其他国家公园的发展经验，对我国今后国家公园的健康持续发展具有重要意义。

At this stage, the construction of national parks in China is still in its infancy, and it is of great significance to study and learn from the development experience of other national parks in the world for the healthy and sustainable development of national parks in China in the future.

五指山 图摄 / 孟志军

中国国家公园
China's National Parks

什么是国家公园？

世界自然保护联盟，简称 IUCN，是世界上规模最大、历史最悠久的全球性非营利环保机构，也是自然环境保护与可持续发展领域的，作为联合国大会永久观察员的国际组织。

世界自然保护联盟将国家公园定义为大面积自然或近自然区域，用以保护大尺度生态过程以及这一区域的物种和生态系统特征，同时提供与其环境和文化相容的精神、科学、教育、休闲和游憩的机会。

International Union for Conservation of Nature, or IUCN for short, is the world's largest and oldest global non-profit environmental protection agency, as well as an international organization in the field of natural environmental protection and sustainable development as a permanent observer to the United Nations General Assembly.
IUCN defines national parks as large natural or near natural areas set aside to protect large-scale ecological processes, along with the complement of species and ecosystems characteristic of the area, which also provide a foundation for environmentally and culturally compatible spiritual, scientific, educational, recreational and visitor opportunities.

世界自然保护联盟的自然保护地分类体系是目前世界上应用最为广泛、接受度最高的自然保护地分类模式，为全球自然保护地提供了一个规划、建立和管理的重要全球标准。IUCN 按照生物多样性的重要性、受人类干扰程度、受保护的严格程度等，将全球保护区分为以下几个类别。

The IUCN protected area management categories are currently the most widely applied and accepted classification model for protected areas worldwide. They provide a crucial global standard for the planning, establishment, and management of protected areas across the globe. IUCN categorizes global protected areas based on criteria such as the importance of biodiversity, the degree of human disturbance, and the level of protection.

Ⅰ - 严格的自然保护地 / 荒野地
Strict Nature Reserve/ Wilderness Area

Ⅱ - 国家公园
National Park

Ⅲ - 自然历史遗迹或地貌
Natural Monument or Feature

Ⅳ - 生境 / 物种管理区
Habitat/Species Management Area

Ⅴ - 陆地景观 / 海洋景观
Protected Landscape/Seascape

Ⅵ - 可持续利用自然资源的保护地
Protected Area with Sustainable Use of Natural Resources

What is National Park ?

中国将国家公园定义为由国家批准设立并主导管理,边界清晰,以保护具有国家代表性的大面积自然生态系统为主要目的,实现自然资源科学保护和合理利用的特定陆地或海洋区域。

National parks, as defined by China, refer to a specific land or marine area that is approved and predominantly managed by the state, with clear boundaries, and its primary purpose is to protect large-scale representative natural ecosystems of the nation, achieving the dual goals of scientific conservation and sustainable utilization of natural resources.

(中共中央办公厅 国务院办公厅)

(The General Office of the CPC Central Committee and the General Office of the State Council)

中国按照自然生态系统原真性、整体性、系统性及其内在规律,依据管理目标与效能并借鉴国际经验,将自然保护地按生态价值和保护强度高低依次分为 3 类:国家公园、自然保护区和自然公园。

China categorizes nature reserves into three classes based on the authenticity, integrity, and systematic of the natural ecosystems, as well as their inherent laws. This classification takes into account management objectives and effectiveness, drawing on international experience, and ranks the reserves according to the ecological value and level of protection intensity.

国家公园生态价值四"最"
Ecological Values of National Parks: The Four "Mosts"

自然生态系统最重要
自然景观最独特
自然遗产最精华
生物多样性最富集

Most Important Natural Ecosystems
Most Unique Natural Landscapes
Most Essential Natural Heritage
Most Abundant Biodiversity

国家公园保护度强
Strong degree of protection of National Parks

保护范围大
生态过程完整
具有全球价值
国民认同度高

Extensive Conservation Coverage
Intact Ecological Processes
Global Significance
High Public Recognition

国家公园

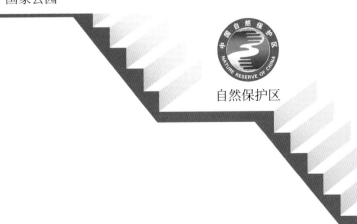

自然保护区

自然公园

走中国的国家公园之路

2013年

中国首次提出建立国家公园体制。

The first proposal to establish a national park system in China is advanced.

2015年

5月，国家发展改革委员会等13部委出台《建立国家公园体制试点方案》。6月，我国启动了为期三年的国家公园体制试点工作。

In May, 13 ministries and commissions, including the National Development and Reform Commission (NDRC), issued the *Pilot Program for the Establishment of the National Park System*. In June, China launched a three-year pilot program for the national park system.

2015年

9月，国务院印发《生态文明体制改革总体方案》，要求加强对国家公园试点的指导，在试点基础上研究制定建立国家公园体制总体方案，构建保护珍稀野生动植物的长效机制。

In September, the State Council issued the *Overall Plan for the Reform of the Ecological Civilization System*, calling for the strengthening of guidance for pilot national parks and the study and formulation of an overall plan for the establishment of the national park system on the basis of the pilots. A long-term mechanism for the protection of rare wildlife is to be constructed.

Taking China's Path to National Parks

中共中央办公厅、国务院办公厅印发《建立国家公园体制总体方案》。

The General Office of the CPC Central Committee and the General Office of the State Council issued the *Overall Plan for the Establishment of the National Park System*.

2017 年

中共中央办公厅、国务院办公厅印发了《关于建立以国家公园为主体的自然保护地体系的指导意见》，要求构建科学合理的自然保护地体系；建立统一规范高效的管理体制；创新自然保护地建设发展机制；加强自然保护地生态环境监督考核。

The General Office of the CPC Central Committee and the General Office of the State Council issued the *Guiding Opinions on the Establishment of a Nature Reserve System with National Parks as the Mainstay*, which demands to build a scientific and rational nature reserve system; to establish a unified, standardized and efficient management system; to innovate the mechanism for the construction and development of nature reserves; to strengthen the ecological environment supervision and assessment of nature reserves.

2019 年

《中共中央 国务院关于全面推进美丽中国建设的意见》发布，强调全面推进以国家公园为主体的自然保护地体系建设，到 2035 年，国家公园体系基本建成，筑牢自然生态屏障，生态系统多样性稳定性持续性显著提升。

State Council issues *Opinions on Comprehensively Promoting the Construction of the Beautiful China*, emphasizing the comprehensive promotion of the construction of a system of nature reserves with national parks as the main body, and the basic completion of the national park system by 2035, building a solid natural ecological barrier and enhancing the diversity, stability and continuity of ecosystem.

2023 年

帕罗山 图摄 / 孟志军

第一批中国国家公园　The First China National Parks

　　2021 年 10 月 12 日，国家主席习近平在《生物多样性公约》第十五次缔约方大会领导人峰会的主旨讲话中宣布，中国正式设立三江源、大熊猫、东北虎豹、海南热带雨林、武夷山等第一批国家公园。让我们一起来认识中国国家公园的 5 位成员吧！

On October 12th, 2021, President Xi Jinping attended the 15th Conference of the Parties to the *Convention on Biological Diversity* (COP15) and delivered a keynote speech via video. In his speech, President Xi announced the official establishment of the first batch of national parks in China, including Three-River-Source, Giant Panda, Northeast China Tiger and Leopard, Hainan Tropical Rainforest, and Mount Wuyi. Let's get to know these five members of China National Parks together!

咱家标识是由海南名山五指山和海南热带雨林旗舰物种海南长臂猿融合而成，寓意为以海南长臂猿为代表的热带雨林动植物生生不息！

Our logo is a fusion of Hainan's famous Wuzhi Mountain and the flagship species of Hainan's tropical rainforest, the Hainan gibbon, implying that the tropical rainforest flora and fauna represented by the Hainan gibbon are everlasting!

三江源国家公园

东北虎豹国家公园

大熊猫国家公园

武夷山国家公园

海南热带雨林国家公园

位置：海南岛中部山区
面积：4269 平方千米
特征：我国分布最集中、类型最多样、保存最完好、连片面积最大的大陆性岛屿型热带雨林，是热带生物多样性和遗传资源的宝库
旗舰物种：海南长臂猿

Location: The central mountainous area of Hainan
Area: 4269 km²
Feature: The continental island-type tropical rainforest with the most concentrated distribution, the most diverse types, the best preservation and the largest contiguous area in China, which is a treasure trove of tropical biodiversity and genetic resources
Flagship species: Hainan gibbon

五指山 图摄/孟志军

第三章
探秘海南热带雨林国家公园

吊罗山 图摄/张哲

毛瑞 图摄/孟志军

Chapter 3
Exploring the National Park of Hainan Tropical Rainforest

海南热带雨林国家公园的显著特点

The Distinguished Features of the National Park of Hainan Tropical Rainforest

尖峰岭日出 图摄 / 孟志军

热带雨林分布最集中

Most Concentrated Distribution of Tropical Rainforests

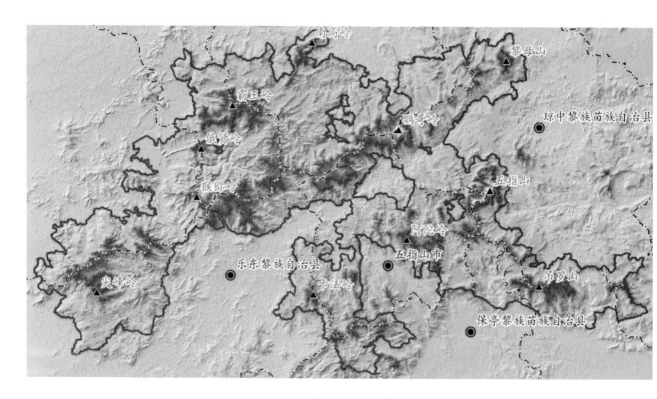

海南热带雨林国家公园范围示意图
图源：《海南热带雨林国家公园规划（2019－2025年）》

  海南热带雨林公园集中分布在海南岛东南、中南、西南部和中部海拔500米以上的山地，包括鹦哥岭、霸王岭、尖峰岭、黎母山、吊罗山、五指山、毛瑞7个山区，涉及海南省中部的五指山、琼中、白沙、东方、陵水、昌江、乐东、保亭、万宁9市县。其间和周边还集中分布着黎族、苗族等传统少数民族聚落（或村落），其中黎族是海南的世居少数民族。

National Park of Hainan Tropical Rainforest is predominantly distributed in the southeastern, central-southern, southwestern, and mountainous central regions of Hainan Island, including seven mountainous areas: Yinggeling, Bawangling, Jianfengling, Limu Mountain, Diaoluo Mountain, Wuzhi Mountain, and Maorui. The coverage extends to nine cities and counties in the central part of Hainan Province, namely Wuzhishan, Qiongzhong, Baisha, Dongfang, Lingshui, Changjiang, Ledong, Baoting, and Wanning. In and around these areas, there is also a concentration of traditional Li and Miao enthnic minority settlements (or villages), with the Li being a long-standing minority group in Hainan.

物种类型最多样
Greatest Variety of Species Types

海南热带雨林国家公园物种统计
Species Census of National Park of Hainan Tropical Rainforest

Bryophytes	苔藓植物	79科 209属 714种
Ferns	蕨类植物	32科 122属 522种
Gymnosperms	裸子植物	6科 10属 26种
Angiosperms	被子植物	172科 1027属 3105种
Mammals	哺乳类	9目 26科 55属 85种
Birds	鸟类	19目 71科 195属 305种
Reptiles	爬行类	3目 24科 63属 95种
Amphibians	两栖类	2目 8科 28属 48种
Fish	鱼类	7目 23科 88属 118种

高等植物 289科 1368属 4367种
国家一级保护野生植物 6种
国家二级保护野生植物 143种
中国特有植物 427种
海南特有植物 419种

野生脊椎动物 651种
国家一级保护野生动物 14种
国家二级保护野生动物 131种
海南特有野生动物 23种

Higher plants: 289 families, 1368 genera, and 4367 species.
National First-Class Protected Wild Plants: 6 species.
National Second-Class Protected Wild Plants: 143 species.
China's endemic plants: 427 species.
Hainan's Unique Plants: 419 species.

Wild Vertebrates: 651 species.
National First-Class Protected Wild Animals: 14 species.
National Second-Class Protected Wild Animals: 131 species.
Hainan's Unique Wild Animals: 23 species.

数据来源：海南热带雨林国家公园官网

陆均松
Dacrydium pectinatum de Laub.
图摄 / 孟志军

海南拟髭蟾
Hainan Pseudomoustache Toad
图摄 / 陈枳衡

霸王岭睑虎
Bawangling Tiger Gecko
图摄 / 张哲

海南长臂猿
Hainan gibbon
图摄 / 范朋飞

海南热带雨林国家公园仅占我国陆地国土面积的万分之四，孕育的维管植物、两栖动物、爬行动物、鸟类和哺乳动物种类的比例却均位于中国前列。这里还是中国特有、热带特有、世界特有的动植物种类及种质基因库。

National Park of Hainan Tropical Rainforest, covering four-thousandths of China's land area, harbors a high proportion of vascular plants, amphibians, reptiles, birds, and mammals, all ranking among the top in China. What's more, this area is home to unique, tropical, and globally exclusive plant and animal species, serving as a genetic repository for biodiversity in China and the world.

资源保存最完好
Best-Preserved Resources

原始制陶 primitive pottery

黎锦 Li brocade

雨林中的黎族船型屋 boat-shaped houses of the Li ethnic group in the rainforest

图摄 / 孟志军

海南热带雨林不仅保存着较为完整的植被垂直带谱，还为山区中的黎族和苗族等少数民族的繁衍生息提供了必要的物质基础和生产条件，保存了"雨林民族"众多历史悠久的古迹遗址和瑰丽多元的民族风情。

国内热带雨林连片面积最大

Largest Contiguous Area of Tropical Rainforests in China

树皮衣
bark clothes

霸王岭五指神树
Five Fingers of the God Tree in Bawangling

图源：海南热带雨林国家公园官网

海南热带雨林国家公园区划总面积为 4269 平方千米，约占海南岛陆域面积的 12.1%。森林面积为 4207.68 平方千米，占中国热带雨林面积的近三分之一。

The National Park of Hainan Tropical Rainforest covers a total area of 4269 km², accounting for about 12.1% of the land area of Hainan Island. The forest area is 4207.68 km² in total, accounting for nearly one-third of China's tropical rainforest area.

图摄 / 黎夏壮

Hainan tropical rainforest not only preserves a relatively intact vertical vegetation spectrum but also provides the necessary material basis and production conditions for the reproduction and livelihoods of ethnic groups such as the Li and Miao in mountainous areas. It also safeguards numerous historic and culturally rich heritage sites of the "rainforest ethnic groups" and their diverse and vibrant ethnic traditions.

海南热带雨林国家公园总体规划
（2022—2030 年）

2025 年   **2030 年**

野生海南长臂猿种群数量

母长臂猿 海南省林业局

达到 **38** 只以上
Increase the wild Hainan gibbons' population to 38 or more.

达到 **50** 只以上
Increase the wild Hainan gibbons' population to 50 or more.

人工林退出商品性经营面积

达到 **300** 平方千米
Retire 300 square kilometers of artificially managed forests from commercial operations.

达到 **500** 平方千米
Retire 500 square kilometers of artificially managed forests from commercial operations.

五指山 图摄 / 孟志军

完成热带雨林类型与结构、重点保护动植物本底资源调查
Complete surveys of tropical rainforest types, structures, and key animal and plant resources.

自然教育受众达到 **300** 万人次
Reach an audience of 3 million people for nature education.

"天空地"一体化监测体系覆盖率达到 **60%**
Achieve a 60% coverage rate for an integrated "sky-to-ground" monitoring system.

主要水体Ⅰ类和Ⅱ类水质比例达到 **98%**
Reach a proportion of 98% for Class I and Class II water quality grades in major water bodies.

管护站点建设完成率达到 **90%**
Accomplish the construction of protection and management stations with a completion rate of 90%.

吊罗山 图摄 / 孟志军

The National Park of Hainan Tropical Rainforest Master Plan (2022—2030)

海南热带雨林国家公园的建设，推动了海南热带雨林生态系统多样性、稳定性、持续性不断提升，人与自然和谐共生的愿景正逐步实现。这里承载着无数生灵的生息、繁衍与传承，也承载着人类亲近自然、探索自然、保护自然的期待。希望在国家规划之下，海南热带雨林国家公园能成为热带雨林珍贵自然资源传承和生物多样性保护的典范。

The construction of the National Park of Hainan Tropical Rainforest has propelled the continuous enhancement of diversity, stability, and sustainability in the Hainan tropical rainforest ecosystem. The vision of harmonious coexistence between humans and nature is gradually becoming a reality. Here, it carries the livelihood, reproduction, and inheritance of countless creatures, as well as the expectations of humanity to approach, explore, and protect nature. With the national planning, it is hoped that the National Park of Hainan Tropical Rainforest can become an exemplary model for the inheritance of precious natural resources and the conservation of biodiversity in tropical rainforests.

五指山 图摄/孟志军

热带雨林秘境有奇观

　　原始是海南热带雨林的真实样子，板根、空中花园、植物绞杀、老茎生花、独木成林、滴水叶尖、巨藤飞舞、根抱石八大奇观也是热带雨林区别于其他森林生态的典型特征！

The primitive is the authentic appearance of Hainan's tropical rainforest. The eight wonders, including buttress roots, aerial gardens, strangler figs, cauliflory, one tree makes a forest, dripping-tip leaves, giant vines dancing, and roots embracing rocks, are also the signature features that distinguish tropical rainforests from other forest ecosystems!

Abundance of Wonders in the Tropical Rainforest Secrets

雨林秘境 图摄/孟志军

板根

Buttress Roots

穿梭在热带雨林中,你可能会被一块巨大的"板"绊倒,这个就叫板根。它们宽大、坚固,就像一块块扎实的木板。板根是热带雨林中树木生存的秘密武器之一,它们犹如板状支架将树木支撑起来,增加了树木的稳定性和抵抗风力的能力,帮助树木在湿润的土壤和不稳定的地形上存活。板根的高度可达10米,可向外延伸10多米,尤为壮观。

Traveling through the rainforest, you may trip over a huge "board" called buttress roots. It gets its name from its broad, solid shape, which resembles a solid plank of wood. It is one of the secret weapons for the survival of trees in tropical rainforests. The solid plate root acts as a plate-like support for the tree, increasing its stability and resistance to wind, and helping it to survive in wet soils and unstable terrain. Buttress roots are particularly spectacular as they can reach up to 10 meters in height and can extend more than 10 meters outward.

板根 图摄/张哲

空中花园
Aerial Gardens

在热带雨林，植物对于生存空间的竞争十分激烈，许多植物都在寻找高处的生存空间。雨林里的大树上长满了各种各样的蕨类、地衣、兰科植物等，形成了一个个"空中花园"。

In tropical rainforests, plants compete fiercely for living space, and many plants are looking for living space in high places. Large trees in the rainforest are covered with a variety of ferns, lichens, orchids and other plants, forming "Aerial gardens".

空中花园 图摄/张哲

植物绞杀
Strangler Figs

绞杀是植物间争夺生存空间的一种方式，但这种方式是极为残酷的，在无声中完成杀戮。"杀手"一般是由谁来扮演呢？这里给大家介绍一下这个"杀手大家族"——桑科榕属植物，它们通常善于攀爬，在高大的树木身上蔓延寄生，紧紧缠绕树干抑制其生长，阻止其输送水分和养分，并在树冠之上野蛮扩张，掠夺阳光，最终寄主会在绞杀植物的怀中枯萎、腐烂。

Strangulation is a way of fighting for survival among plants, but this way is the most brutal, where the killing is done in silence. Who usually plays the role of the "killer"? Here to introduce you to this "big family of killers" Ficus (Moraceae) plant. They are usually good at climbing, spreading parasitism on tall trees, tightly wrapping around the trunk to inhibit its growth, preventing it from transporting water and nutrients, and expanding savagely above the canopy, robbing the sunlight, and ultimately the host will wither and rot in the arms of the strangler figs.

植物绞杀　图摄/张哲

老茎生花

Cauliflory

大家都知道，大多数植物的花朵都需要昆虫为其授粉才能结出种子，繁衍生息。但在热带雨林中，成年乔木的枝叶通常很高，如果花开在很高的地方，授粉者可能看不到、够不着。因此，为了让昆虫等授粉者注意到花朵，获得更多的授粉机会，树木会将花朵开在接近地面的树干上，从而就形成了老茎生花的景观。

As we all know, most plants need insects to pollinate their flowers in order to produce seeds and reproduce. However, in tropical rainforests, the branches of adult trees are usually very high. If the flowers bloom in a very high place, pollinators may not see or can not reach them. Therefore, in order to be noticed by insects and other pollinators and get more pollination opportunities, the trees will bloom in the trunk of the tree close to the ground, forming a cauliflory landscape.

老茎生花 图摄/张哲

独木成林

海南鹦哥岭独木成林 海南省林业局供图

One Tree Makes a Forest

独木难成林，但榕树却做到了！它能够独立形成一片森林。榕树枝叶茂盛、树冠巨大，枝条上生长的气生根向下延伸进入土壤便会形成新的树干，这些根被称为"支柱根"。榕树高达30米，可向四面伸展，其"支柱根"和枝干相互交织，进而形成一片独立的森林，是热带雨林中的一大景象。

It is difficult to make a forest out of a single tree, but the banyan tree does! It is able to form a forest on its own. The banyan tree has a huge canopy of leaves and branches that grow aerial roots that extend down into the soil and form new trunks, known as "strut roots". The banyan tree reaches 30 meters in height and spreads in all directions, with its "pillar roots" and branches intertwining to form a separate forest.

根抱石

热带雨林中的鸟类或者其他外力将植物的种子带到裸露的岩石缝里，种子生根发芽，根系包裹着岩石外围不断生长，支撑起高大的树干和宽广的树冠，从而形成了"根抱石"的奇观。根抱石也可能由榕树的气生根绕岩石生长而成。从某种角度来看，这也可以看作是植物对岩石的"绞杀"。

In tropical rainforests, birds or other external forces carry plant seeds into exposed rock crevices. These seeds take root, and the roots gradually wrap around the periphery of the rocks, growing continuously. This process supports the tall trunks and expansive canopies, creating the spectacle known as "roots embracing rocks". It can also be formed by aerial roots of banyan trees growing around rocks. From a certain perspective, it can be seen as plants "strangling" the rocks.

滴水叶尖 图摄 / 张哲

滴水叶尖
Dripping-tip Leaves

在热带雨林高温多雨的环境中，为了让叶面尽快变干，不妨碍叶片的蒸腾作用，同时避免一些微小的附生植物在叶面生长进而阻碍光合作用，雨林植物叶片演变出"滴水叶尖"，可以让叶片表面的雨水聚集到叶尖，变成水滴顺叶尖流下。

In the hot and rainy conditions of the tropical rainforest, to facilitate the rapid drying of leaf surfaces without hindering transpiration and to prevent the growth of small epiphytic plants on the leaf surface that may impede photosynthesis, rainforest plants have evolved "dripping-tip leaves". This feature allows rainwater on the leaf surface to gather at the leaf tip, forming droplets that then fall away.

Roots Embracing Rocks

巨藤 飞舞 图摄 / 李清雪

巨藤飞舞
Giant Vines Dancing

在海南热带雨林，常能见到犹如"过江龙"的巨大榼藤，为了争夺更靠近阳光的空间，它们或是攀附大树而上，或是靠着峭壁、山石生长。从远处看，巨藤横亘树间，缠缠绕绕，甚是壮观。

In the tropical rainforests of Hainan, you can often see enormous vines resembling a "flying dragon crossing the river". In a competition for sunlight, they either climb up large trees or grow alongside cliffs and rocks. When viewed from a distance, they span between trees, entwining in a spectacular display.

五指山根抱石 图摄 / 姜恩宇 林业局供图

猿声深处万物生

Untouched Wilderness Bawangling

霸王岭 图摄/孟志军

霸王岭"原住民"——海南长臂猿

The Indigenous Residents of Bawangling - Hainan Gibbons

你知道海南长臂猿在海南住了多少年吗?
Do you know how many years Hainan Gibbons have been living in Hainan?

海南长臂猿已经在海南居住了一万年以上,是世界上唯一一种群数量不到100只的灵长类动物,濒危程度甚至比"国宝"大熊猫要高得多!

Hainan Gibbons have been residing in Hainan for over 10000 years. They are the only primate species in the world with a population of less than 100 individuals, and their level of endangerment is even higher than that of the "national treasure" panda!

海南长臂猿 林业局供图

这些"原住民"可不一般,每一只都非常有个性!不信?一起来看看!

These indigenous residents are not ordinary. Each one of them has a distinct personality! Don't believe it? Let's take a look together!

美食家：海南长臂猿是真正的"吃货"！它们的食谱非常多元化，有超过 140 种不同的食材，雨林中酸酸的野生荔枝是它们的最爱。每年五六月荔枝熟了，就会看到海南长臂猿停留在树旁，"日啖荔枝三百颗"。但如果水果少了，它们也会吃嫩嫩的树叶、花朵，甚至一些小小的昆虫，比如飞蚁、虫蛹、蜘蛛等。

Gourmets: Hainan gibbons are true food lovers! Their diet is incredibly diverse, with over 140 different food items. They particularly favor the sour wild lychee in the rainforest. Every year, in May and June, when lychees ripen, we can see Hainan gibbons lingering by the tree, "eating three hundred lychees a day".But if fruits are scarce, they'll also consume tender leaves, flowers, and even small insects like flying ants, pupae, spiders, and so on.

高空飞行"猿"：海南长臂猿不爱下地走路，它们每天穿梭在树林之间，像一群高空飞行员！就算口渴了，它们也不下地找水，而是在树洞里和大树叶上找积存的雨水或露水来喝。

High-Flying"Apes": Hainan gibbons don't like to walk on the ground. They shuttle through the trees every day, resembling a group of high-flying pilots! Even when thirsty, they don't come down to find water. Instead, they drink from accumulated rainwater or dew found on the tree hollows and large leaves.

自由探险家：海南长臂猿可是随性派，它们不喜欢建房子，也没有特定的地方睡觉。每当太阳下山的时候，它们就会选择一棵大树或是有藤蔓的树作为家，就像是一帮探险家。

Free Explorers： Hainan gibbons are quite carefree. They don't like building houses and have no specific sleeping spots. When the sun goes down, they simply choose a big tree or trees with vines as their home. They just like a group of adventurers.

海南长臂猿 图摄 / 钟旭凯

时尚达人：海南长臂猿们还很时尚！雌性海南长臂猿的长相比较温柔，身着金黄色外衣，头顶有一簇小黑点。而雄性海南长臂猿则身着全黑外衣，头顶的黑发蓬松，像个时尚偶像。

Fashion Enthusiasts: Hinan gibbons are very fashionable! The appearance of female Hinan gibbons is rather gentle, with a full coat of gold and a cluster of black dots on top of their head. On the other hand, male Hinan gibbons have an all-black coat, and the black hair on their heads is fluffy, resembling a fashion idol.

一对海南长臂猿 林业局供图

海南明星动物知多少

海南坡鹿

Hainan Eld's Deer

海南坡鹿是海南特有的珍稀物种,被列为国家一级保护野生动物。坡鹿是印度泽鹿的同属,外形与梅花鹿相似,但体形较小,花斑较少。由于它分布在海南西部的丘陵草坡地带,故称海南坡鹿。

The Hainan Eld's deer is a rare and unique species exclusive to Hainan, listed as a national first-class protected wildlife animal. Belonging to the same genus as the Indian swamp deer, it bears a resemblance to the sika deer but is smaller in size and has fewer spots. Due to its habitat in the hilly grassy slopes of western Hainan, it is called the Hainan Eld's deer.

海南坡鹿 图摄/张哲

睑虎

睑虎,也叫亚洲壁虎,它和我们日常看见的壁虎很不一样,区别在于大部分壁虎是无法闭眼,而睑虎的眼睑是可以活动的,所以它们会眨眼哦!海南睑虎、霸王岭睑虎、周氏睑虎、中华睑虎、光华睑虎均仅分布于海南岛。

How Much Do You Know About Hainan's Star Species?

Goniurosaurus

Goniurosaurus, also known as Asian geckos, are very different from the geckos we see every day, the difference is that most geckos can't close their eyes, while the eyelids of the *Goniurosaurus* can be active, so they will wink! The *Goniurosaurus hainanensis*, the *Goniurosaurus bawanglingensis*, the *Goniurosaurus zhoui*, the *Goniurosaurus sinensis*, and the *Goniurosaurus kwanghua* are all distributed only on Hainan Island.

睑虎 图摄 / 张哲

丽拟丝蟌 *Pseudolestes mirabilis* (Kirby, 1900)

丽拟丝蟌是海南特有物种，也是拟丝蟌科的唯一代表物种。它们头部和面部为蓝色，后翅明显短于前翅，前翅透明，而后翅颜色艳丽——雄性具有金黄色斑，雌性则具有紫红色金属光泽。如此美丽的它，可以说是海南生物中的"颜值担当"。

Pseudolestes mirabilis is endemic to Hainan and the only representative of the Pseudolestidae. They have a blue head and face, the hind wings are obviously shorter than the forewings, the forewings are transparent, and the hindwings are brilliantly colored - the males have golden-yellow spots, while the females have a purplish-red metallic luster. It is so beautiful that it can be called the knockout of Hainan's creatures.

你还知道海南热带雨林的哪些明星动物吗？一起来分享吧！

Do you know which star animals inhabit the tropical rainforests of Hainan?
Let's share together!

丽拟丝蟌 图摄 / 张哲

海南热带雨林的宝藏植物

白桫椤 图摄/陈枳衡

活化石——桫椤
Living Fossil - *Alsophila spinulosa*

如果你行走在海南热带雨林，一路上随着海拔攀升，你会看到一株株高大的蕨类植物，这是被人们称为"活化石"的桫椤，也是世界上现存非常古老的树种之一，毕竟它可是侏罗纪时期恐龙的主要食物呢！

If you walk in the Hainan tropical rainforest, some tall ferns will come into your eyes as the altitude climbs along the way, which is known as the "living fossil"—*Alsophila spinulosa*, one of the world's oldest surviving tree species, which was the main food of dinosaurs in the Jurassic period.

桫椤是目前已经发现的唯一的木本蕨类植物，极其珍贵。海南拥有我国桫椤科的全部属和亚属，表现出系统分类上的完整性。

Alsophila spinulosa is the only woody fern that has been discovered so far, which is extremely precious. Hainan has all the genera and subgenera of the family Cyatheaceae in our country, showing the completeness of the systematic classification.

Exploring the Treasure Rainforest

雨林"神树"——陆均松
"God Tree" in Rainforests
Dacrydium pectinatum de Laub.

要说谁能在热带雨林的林木界里"称王",陆均松可争得一席之位。在海南吊罗山有一棵 2600 多年树龄的陆均松,它的树干需要十几个人才能抱住,被称为"神树"或"树王"。成年的陆均松一般高 20 米以上,胸径也在 100 厘米以上,树龄可达上千年。它是海南山地热带雨林的代表树种之一,是构建天然群落的关键树种,也是顶级植物群落的建群树种。

When it comes to who can be the "king" in the world of tropical rainforest trees, *Dacrydium pectinatum* de Laub. can contend for a prominent position. In Hainan, there is a 2600-year-old *Dacrydium pectinatum* de Laub. in Diaoluo mountain, its tree trunk needs more than a dozen people to surround, known as the "God Tree" or "King of Trees". Adult *Dacrydium pectinatum* de Laub. generally reach a height of over 20 meters, with a diameter at breast height exceeding 100 centimeters, and their age can be over a thousand years. It is one of the representative tree species in Hainan's mountainous tropical rainforests, a key species in constructing natural communities, and a foundational tree species for top-level plant communities.

陆均松 图摄 / 孟志军

鹦哥岭飞瀑草 图摄 / 卢刚

雨林新同学——鹦哥岭飞瀑草
Rainforest New Classmate
Cladopus yinggelingensis

2016 年,研究人员在鹦哥岭片区内白沙黎族自治县南开乡高峰村委会道银村附近发现了两个川苔草科植物新种,其中一个就是鹦哥岭飞瀑草,属于国家二级保护野生植物。它们生活在水质清澈、洁净的环境中,在湍急的河床石头上生长。紧凑生长的飞瀑草开着可爱的小花,随波摇曳,并将花粉包在水泡里传播授粉。

In 2016, researchers discovered two new species of plants near Daoyin Village, Nankai Township, Baisha Li Autonomous County, within the Yinggeling Protection Area. One of them is *Cladopus yinggelingensis*, which belongs to the national second-class protected wildlife plants. They inhabit environments with clear and clean water, growing tightly on the turbulent riverbed stones. The charming little flowers of *Cladopus yinggelingensis* sway with the waves, and their pollen is encapsulated in water bubbles to facilitate pollination.

峻岭飞瀑山水秀

五指山 图摄/孟志军

Steep Mountains and Splendid Waterfalls

石啃瀑布 图摄 / 张晋

仙石安林 图摄 / 孟志军

"海南屋脊"五指山

Wuzhi Mountain, The Roof of Hainan

  五指山位于海南岛中部，横跨五指山市、琼中县，山岭"五峰如指翠相连，撑起炎荒半壁天"（明·丘浚《五指参天》）。

  五指山最高海拔为 1867 米，比五岳之首的泰山还要高出 300 多米，被誉为"海南屋脊"，是海南主要河流万泉河、昌化江的发源地。

Wuzhi Mountain is located in the central part of Hainan Island, spanning Wuzhishan City and Qiongzhong County. The mountain ridge is described as "five peaks resembling connected fingers, supporting half of the vast sky in the Yan and Huang period" (Ming Dynasty's Qiu Jun, "Wuzhi Cantian"). The highest peak of Wuzhi Mountain is 1867 meters above sea level, more than 300 meters higher than Mount Tai. It is acclaimed as the "Roof of Hainan" and serves as the headwaters of the major rivers, Wanquan River and Changhua River, in Hainan.

  凭借高度优势，这里拥有目前海南岛海拔跨度最大、最为齐全的森林植被类型，共有热带低地雨林、热带山地雨林、热带亚高山矮林、热带山顶灌丛和次生热带雨林 5 种森林植被类型，植被垂直带谱完整，原始完整地保留了 5 个海拔垂直高度的生物多样性。

With the height advantage, this place has the largest altitude span and the most complete type of forest vegetation in Hainan Island, including a total of tropical lowland rainforest, tropical mountain rainforest, tropical subalpine dwarf forest, tropical hilltop scrub and secondary tropical rainforest of five types of forest vegetation. The vegetation vertical zone spectrum is complete, and the original and complete biodiversity of five altitude vegetation is preserved.

五指山 图摄 / 孟志军

"海南第一瀑"——枫果山瀑布
"Hainan's First Waterfall" - Fengguoshan Waterfall

百瀑飞流吊罗山

A Hundred Waterfalls at Diaoluo Mountain

　　吊罗山可不是一般的地方，这里是瀑布的天堂！这里雨水多，水资源非常丰富，山体地形起伏大、变化多，水顺着山体的沟沟壑壑从高处往下流，造就了多河谷、多瀑布的独特景观。

Diaoluo Mountain is truly an extraordinary place. It's a waterfall paradise! It receives abundant rainfall, making water resources very rich. The undulating terrain of the mountain creates numerous changes in elevation, causing water to flow down through the gullies and valleys, resulting in its characteristic of having many valleys and waterfalls.

　　吊罗山的瀑布景色非常壮美，在海南热带雨林国家公园中独具特色。吊罗山不仅有"海南第一瀑"——枫果山瀑布，还有"网红瀑布"——大里瀑布、石晴瀑布和未起名大小瀑布上百个，享有"梦幻雨林，百瀑吊罗""吊罗归来不看水"的美誉。

The waterfall scenery at Diaoluo Mountain is extraordinarily magnificent and is unique in Hainan Tropical Rainforest National Park. Diaoluo Mountain is home to "Hainan's First Waterfall"—Fengguoshan Waterfall, as well as popular waterfalls such as Dali Waterfall, Shiqing Waterfall, and many smaller unnamed waterfalls, totaling over a hundred. It has earned itself the reputation of "Dreamy rainforest, hundred waterfalls at Diaoluo"and "Back from Diaoluo, no longer gazing at waters".

南喜瀑布 图摄 / 张哲

雨林也有喀斯特！
Karst Landforms in Rainforests

海南拥有壮观的喀斯特地貌。这就不得不提到海南最大的石灰岩山体——俄贤岭，它有着海南现存面积最大、原始状态保存最为完整、从低海拔到高海拔的喀斯特地貌原始热带雨林。放眼望去，怪石嶙峋的山峰秀丽挺拔，不愧是"巨石上的热带雨林"。

Hainan is blessed with spectacular karst landscapes. This brings us to the largest limestone mountain in Hainan, the Exianling, which has the largest area and the most well-preserved original state of the karst landscape from low altitude to high altitude Hainan's original tropical rainforest. As far as the eye can see, the jagged peaks are beautiful and upright, worthy of the title "the rainforest on boulders".

俄贤岭 图摄/孟志军

仙安石林 图摄/孟志军

　　毛瑞的"仙安石林"是大自然的绝世奇观,被誉为喀斯特地貌中的"稀世珍宝"。这种热带雨林型的喀斯特针状石林在世界上仅发现两例,它们看起来就像魔法王国里的巨大石柱森林!

Maorui's "Xian'an Stone Forest" is one of nature's greatest wonders, known as a "rear treasure" among karst landforms. This type of tropical rainforest karst needle-like stone forest has only two known examples in the world, like giant pillar forest in the magic world.

　　和其他常见的热带雨林景观相比,"仙安石林"危峰耸立,难以攀行。从远处眺望时,石林像狼牙一样密集;走到近处看,可以看见层层叠叠的石头,千岩万壑。每一处景观都在告诉人们,它就是海南热带雨林国家公园中最独特的宝物!

Compared to other common tropical rainforest landscapes, the Xian'an Stone Forest stands out with its towering and almost impassable peaks. When viewed from a distance or from above, the stone forest appears densely packed like a row of sharp teeth. But upon closer inspection, you can see layer upon layer of rocks forming a landscape of rugged and intricate beauty. Every nook and cranny tells you that it is the most unique gem within Hainan Tropical Rainforest National Park!

仙安石林 图摄/孟志军

73

黎苗人家风情浓

Rich Cultural Charm of Li and Miao Ethnic Groups

黎母山 图摄 / 孟志军

海南最早的居民

根据考古材料，黎族先民早在3000多年前的殷周时就定居于海南，是最早一批来到这里的人。他们生活在海南热带雨林之间、中南部海滨平原之上。早在宋元时期，精美华丽的黎锦就名噪一时。

According to archaeological evidence, the ancestors of the Li ethnic group settled on Hainan Island over 3000 years ago, around the Yin and Zhou periods, making them among the earliest inhabitants of the island. They lived between the tropical rainforests and the coastal plains in the central-southern part of Hainan. As early as the Song and Yuan dynasties, the exquisite and gorgeous Li Brocade gained widespread fame.

海南黎族居住的茅草屋就像一艘艘倒扣的小船。最早的船型屋是干栏式建筑，上层住人，下层饲养牲畜。随着生产力水平的提高，船型屋下方的干栏结构逐渐消失，变成了"金字屋"。

The thatched houses of the Li people in Hainan are like upside-down boats. The earliest boat-shaped houses had a stilt structure, with people living on the upper level and keeping livestock on the lower level. With improved productivity, the stilt structure below gradually disappeared. The boat-shaped house transformed into a "gold-character house".

The Earliest Inhabitants in Hainan

船型屋 图摄 / 张哲 海南省林业局供图

明万历年间，两广总督张鸣冈调派"广西苗兵"等到海南镇压起义，后苗兵定居海南，成为海南的世居少数民族。黎苗两族在此伴生，文化相互融合，共同孕育出独具特色的黎苗文化。在海南省 32 项国家级非物质文化遗产中，与黎苗文化相关的非物质文化遗产就占了近一半。

During the Wanli period of the Ming Dynasty, the Governor of Guangdong and Guangxi, Zhang Minggang, dispatched "Guangxi Miao troops" to suppress uprisings in Hainan. Later, the Miao troops settled in Hainan and became one of the long-standing ethnic minorities on the island. The Li and Miao ethnic groups coexisted on Hainan Island, with their cultures blending together, giving rise to the unique and fascinating Li-Miao culture. Nearly half of the 32 national-level intangible cultural heritage items in Hainan Province are related to the intangible cultural heritage of Li - Miao culture.

黎苗文化
Li-Miao Culture

三月三
The March 3rd Festival

三月三，是黎族人民最盛大的传统节日！在这个节日里，人们会穿上漂亮的节日服装，带着特别的食物，比如山兰米酒和竹筒香饭，从四面八方聚集到一起，或祭拜始祖，或成群结队地唱歌、跳舞、吹奏打击乐器，共同欢庆佳节。年轻的男女们会在这个节日狂欢，他们唱歌、跳舞，借此传递心意。他们在爱情的海洋中沉浸，直到天快亮的时候才离开，相约来年三月三再会。

The March 3rd Festival is the grandest traditional festival of the Li people! During this festival, people dress in beautiful traditional clothing and bring special foods like Shanlan wine and bamboo-tube rice. They come together from all directions to either worship their ancestors or gather in groups, singing, dancing, playing music, and celebrating the festival. Young men and women take the opportunity to revel during this festival, singing, dancing, and expressing their affections. They immerse themselves in the sea of love and don't leave until the early hours of the morning, promising to meet again on March 3rd of the following year.

三月三 图摄 / 孟志军

黎锦
Li Brocade

海南黎锦在黎族发展史中具有"甲骨文"级别的文化符号地位，在3000年的发展变迁中沉淀了黎族妇女对生活的记录、对自然的热爱，以及对本民族发展的情感与思考，具有极高的艺术欣赏价值和人文历史研究价值。

Hainan Li Brocade holds a cultural symbol status akin to "oracle bone inscriptions" in the history of Li ethnic development. Over the course of 3000 years, it has encapsulated the records of Li women's lives, their love for nature, and emotions and reflections on the development of their own ethnic group. It possesses exceptionally high artistic appreciation and value for the humanities and historical research.

黎锦 图摄 / 孟志军

苗绣
Miao Embroidery

海南苗绣和黎锦一样，是海南苗族妇女们指尖上的"无字天书"。苗族的妇女们刺绣从不先描画草图，全凭自己天生的感知力、娴熟的技艺和非凡的记忆力，数着底布上的经纬线挑绣。她们善于观察自然，图案多以山水、花鸟、虫鱼等为主，用鲜艳的绣线在蓝底白花间展现海南苗族鲜明的民族个性和艺术特点。

Similar to Li Brocade, Hainan Miao Embroidery is considered the "silent language" at the fingertips of Miao women in Hainan. Miao women do not sketch any drafts before embroidery; instead, they rely on their innate insight, skilled techniques, and exceptional memory. They embroider by counting the warp and weft threads on the base fabric. Proficient in observing nature, their patterns often depict landscapes, flowers and birds, insects, and fish. Using vibrant threads, they showcase the distinctive ethnic personality and artistic characteristics of the Hainan Miao people within the blue background adorned with white flowers.

苗绣 图摄 / 孟志军

苗族盘皇舞
The Miao's Panhuang Dance

苗族盘皇舞也是五指山市的一颗非遗"明珠"，源自盘皇开天地、造万物的传说，是纪念祖先的传统祭祀舞蹈。盘皇舞没有旋律伴奏，只以锣、鼓击打节奏。舞蹈伊始，需先摆好盘皇、三元神及各神画像，设香案。舞蹈在锣鼓声中按上元、中元、下元的顺序进行，极具原始神秘的色彩，又称"三元舞"。

The Miao Panhuang Dance is also a cultural heritage "pearl" in Wuzhishan city, originating from the legend that Panhuang created the heavens and earth and all things. It is a traditional ceremonial dance to commemorate ancestors. The Panhuang Dance does not have musical accompaniment but is accompanied by the beats of gongs and drums. At the beginning of the dance, images of Panhuang, the three original deities, and various gods are arranged, and an incense table is set. The dance commences in the order of Shangyuan (Upper Realm), Zhongyuan (Middle Realm), and Xiayuan (Lower Realm) amid the sound of gongs and drums. The dance is highly primitive and mysterious, also known as the "Three Realms Dance".

苗族盘皇舞 图摄 / 孟志军

第四章
爱在雨林
Chapter 4
Love In the Rainforest

护林员黎炳新巡护雨林　图摄/陈枳衡

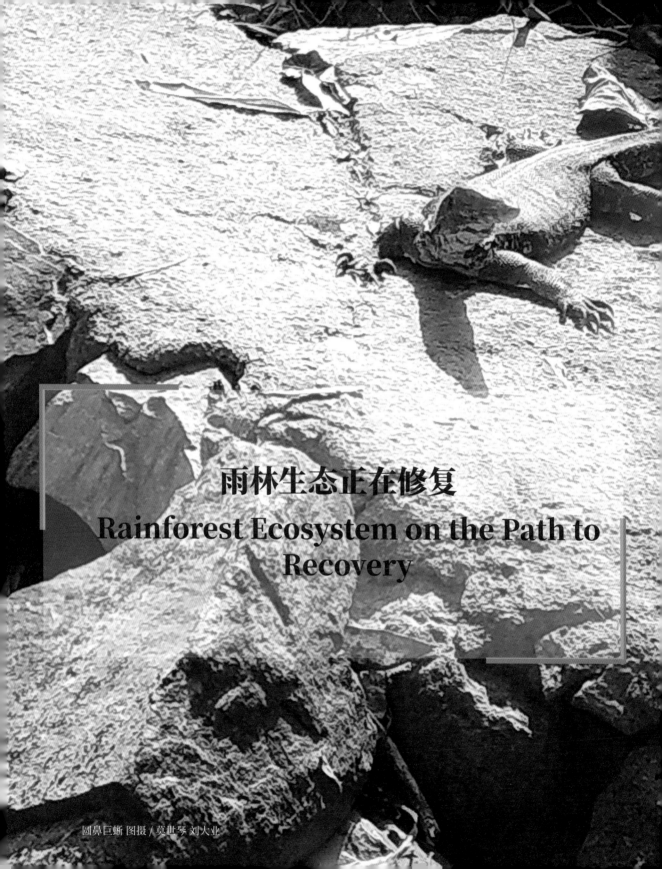

雨林生态正在修复
Rainforest Ecosystem on the Path to Recovery

圆鼻巨蜥 图摄 / 莫世琴 刘大业

圆鼻巨蜥
Varanus salvator

2023 年，海南热带雨林国家公园管理局尖峰岭分局通过红外相机拍摄到了极危物种——圆鼻巨蜥。这也是尖峰岭片区时隔十多年后，第一次记录到这一珍稀野生物种，也是雨林生态正在修复的有力证明。

In 2023, the Hainan Tropical Rainforest National Park Administration's Jianfengling Branch photographed a critically endangered species — *Varanus salvator* through an infrared camera. This is the first recorded sighting of this rare wildlife species in over a decade in the Jianfengling area, serving as strong evidence that the rainforest ecosystem is recovering.

创新体制机制　焕发雨林活力

创新建立扁平化管理体制
Innovation in Establishing a Flat Management System

在海南省林业局加挂海南热带雨林国家公园管理局牌子，整合原有20个自然保护地，联合各地组成多个协调小组，解决了人为割裂、保护空缺等问题。

The Forestry Department of Hainan Province has hung the sign of The National Park of Hainan Tropical Rainforest Management Bureau. It has integrated the original 20 nature reserves, formed a number of coordinating groups with localities, and resolved the problem of man-made fragmentation, protection gaps and other issues.

规范核心保护区生态搬迁
Standardizing Ecological Relocation in core Protection Areas

2020年3月，海南省委、省政府印发《海南热带雨林国家公园生态搬迁方案》，对被划定在海南热带雨林国家公园核心保护区范围内的东方、五指山、保亭、白沙4个市县的11个自然村470户1885人实施生态搬迁。2020年底，作为海南首个实施整村生态搬迁的村庄，白沙黎族自治县高峰村的生活迎来了新的变化。首批乔迁村民搬出大山，通过生产生活条件的改善、绿色特色产业的可持续发展、社会保障的全覆盖，确保村民生活持续向好，收入不断提高。目前，生态搬迁已全部完成，雨林生态更美，人民生活更好。

In March 2020, the Hainan Provincial Party Committee and Provincial Government issued *the National Park of Hainan Tropical Rainforest Ecological Relocation Plan* to carry out ecological relocation work for 470 households of 1885 people in 11 natural villages in four cities and counties, namely Dongfang, Wuzhishan, Baoting and Baisha, which are delineated in the scope of the core protection zone of the National Park of Hainan Tropical Rainforest. At the end of 2020, as the first village to implement the ecological relocation of an entire village in Hainan, life in Gaofeng Village, Baisha Li Autonomous County, ushered in a new change. The first batch of relocated villagers moved out of the mountain through the improvement of production and living conditions, the sustainable development of green speciality industries, and the full coverage of social security to ensure that the villagers' life continues to improve and income continues to increase. At present, the ecological relocation has been fully completed, the rainforest is more beautiful, and people live better!

Innovative Institutional Mechanisms and Revive Rainforest Vitality

创新执法派驻双重管理体制
Innovate the Dual Management System for Law Enforcement Dispatch

林业管理 海南省林业局供图

由森林公安继续承担国家公园区域内的涉林执法工作，实行省公安厅和省林业局双重管理。其他行政执法工作实行属地综合行政执法，并设立国家公园执法大队，派驻到国家公园各分局，统一负责国家公园区域内的综合行政执法工作。

The Forest Public Security Bureau continues to undertake forestry law enforcement within the National Park area, implementing dual management by the Hainan Provincial Public Security Department and the Forestry Department of Hainan Province. Other administrative law enforcement work is carried out by local comprehensive administrative law enforcement, and a National Park Law Enforcement Brigade is established and stationed at various branches of the National Park to uniformly handle comprehensive administrative law enforcement within the National Park area.

推动科研合作国际化
Promoting the Internationalization of Scientific Research Collaboration

海南热带雨林国家公园管理局联合4所重点科研院校组建面向全球开放的海南国家公园研究院，以项目为导向，柔性引进高层次人才及特需人才，与世界自然保护联盟、世界自然基金会等国际机构开展合作。2022年12月，《全球长臂猿联盟保护宣言》正式发布，这是第一个由中国机构主导、联合国际权威组织发起并面向全球发布的物种保护宣言，旨在号召更多科研机构加入保护网络，开展长臂猿保护的国际科研合作与具体工作。

The National Park of Hainan Tropical Rainforest Management Bureau, in collaboration with four key research institutions, has established the Hainan Institute of National Park, open to the global. This institute operates on a project-oriented basis, flexibly attracting high-level and specialized talents. It engages in cooperation with international organizations such as the International Union for Conservation of Nature (IUCN) and the World Wide Fund for Nature (WWF). In December 2022, the *Global Gibbon Network Conservation Declaration* was officially released. This is the first species conservation declaration led by a Chinese organization, initiated by international authoritative organizations, and released globally. Its purpose is to call for more research institutions to join the protection network and engage in international research cooperation and specific actions for gibbon conservation.

在这片曾经破碎的丛林里，海南立大志要修复雨林生态。2021年，海南省委、省政府与国家林业和草原局联合成立了海南热带雨林国家公园建设工作推进领导小组，旨在让自然资源确权登记、资金支持、生态搬迁等工作更高效、顺畅地推进，加强护林。在当时10个国家公园体制试点区中，海南是"第一个吃螃蟹的"。

In this once fragmented jungle, Hainan has set ambitious goals to restore rainforest ecology. In 2021, the Hainan Provincial Committee and Government, in conjunction with the National Forestry and Grassland Administration, jointly established the Hainan Tropical Rainforest National Park Construction Advancement Leadership Group. This collaboration aims to streamline and enhance the efficiency of tasks such as natural resource rights registration, financial support, and ecological relocation, thus strengthening forest conservation efforts. Among the ten national park system pilot areas at that time, Hainan was the first to "take a bite of the crab", leading the way in this endeavor.

护好"国宝",把大山的安宁留给"雨林精灵"

第37只海南长臂猿出生

核心保护区
Core Protection Zones

禁止人为活动
Prohibit Human Activities

管控严格
High Control Level

长期管控
Long Term Control

2024年6月20日召开的国家林业和草原局野生动物保护及国际合作成果新闻发布会上发布,海南长臂猿野外种群数量从40年前的仅存2群不到10只,增长到7群42只。

According to the press conference on the Achievements of Wildlife Protection and International Cooperation held by the National Forestry and Grassland Administration on June 20, 2024, the number of Hainan gibbons in the wild has increased from only 2 groups with fewer than 10 individuals 40 years ago to 7 groups with 42 individuals.

2023年10月12日,"海南长臂猿"线上数字宣传平台正式上线。该平台通过数字化等科技手段,全方位、多角度、多渠道地宣传海南长臂猿的保护成效和科普知识,提升海南长臂猿的知晓度和影响力,营造人人了解、热爱、参与海南长臂猿保护的良好社会氛围。

线上数字平台

On October 12, 2023, the online digital promotion platform for Hainan gibbons was officially launched. Through digital technologies and other means, it comprehensively promotes conservation achievements and popularizes knowledge about Hainan gibbons from various perspectives and through multiple channels. The aim is to increase awareness and influence of Hainan gibbons, fostering a positive social atmosphere where everyone understands, loves, and participates in the protection of Hainan gibbons.

Protecting the "National Treasures" and Preserving the Tranquility of the Mountains for the "Rainforest Fairies"

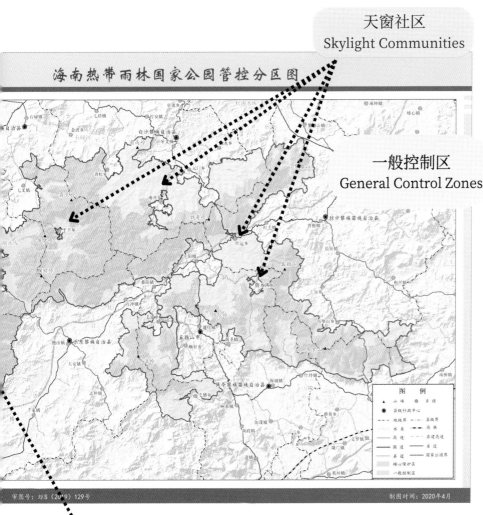

天窗社区
Skylight Communities

一般控制区
General Control Zones

41 个水电站清退或整改

Forty-one hydropower stations within the park will be cleared or rectified.

99% 完成生态搬迁

Ecological relocation completion rate reaches 99%.

24 小时智慧雨林监测

Smart rainforest achieves 24H monitoring.

穿山甲 尖峰岭分局供图

2023年6月，尖峰岭再次发现国家一级重点保护野生动物中华穿山甲的身影，吃白蚁的穿山甲对于保护热带雨林生态系统的健康非常重要！

In June 2023, the presence of the nationally protected animal, the Chinese Pangolin, was once again discovered in Jianfengling. These termite-eating pangolins are crucial for the conservation of the health of the tropical rainforest ecosystem!

逐步恢复区域内生态系统的自然状态，逐步减少人为活动对生态系统的影响。

Gradually restore the natural state of the regional ecosystems and progressively reduce the human activities' impact on the ecosystems.

雨林与您
Rainforest and You

海南热带雨林国家公园在有效保护生态资源的同时也在推进生态产品价值实现，引领社会不同群体亲近雨林、认识雨林、保护雨林，传递着人与雨林和谐共生共发展的理念。

While effectively conserving ecological resources, the National Park of Hainan Tropical Rainforest is also promoting value realization of ecological products. It leads various social groups to approach, understand, and protect the rainforest, conveying the concept of harmonious coexistence and development between humans and the rainforest.

解锁雨林奥秘 Decoding Rainforests Mysteries

从 1957 年开始,位于海拔 800 余米的尖峰岭林区,"大隐于野"的尖峰岭生态站就已经开始了科研监测工作。直到 1986 年,该站点才正式建立,如今已经成为海南热带雨林国家公园设立之初的本底参考数据库。小小的科研站点对科研工作、科普教学、政府管理部门的数据共享和支持服务起到了大大的作用。

科研监测设施 尖峰岭分局、尖峰岭生态定位站供图

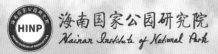

海南国家公园研究院的成立，汇集了国内外 300 余位各领域专家学者，为国家公园建设提供科技支撑、智力支撑。

The establishment of the Hainan Institute of National Park has brought together over 300 experts and scholars from various fields, both domestic and international, providing technology and intellectual support for the development of the national park.

Since 1957, the remote "Jianfengling Ecological Station" located at an altitude of over 800 meters in the Jianfengling forest area has been conducting research and monitoring work. It wasn't until 1986 that the station was formally established. Today, it serves as the baseline reference data for the establishment of the National Park of Hainan Tropical Rainforest. Despite its small size, this research station plays a significant role in scientific research, science education, and government data sharing and support services.

护林员观鸟 海南省林业局供图

洪水村 图摄/孟志军

五指山水满乡茶园 图摄/孟志军

位于"天窗社区"的五指山市水满乡毛纳村，乘着海南热带雨林国家公园建设的东风，加快生态旅游和大叶茶产业发展。2022年4月至2023年8月，毛纳村累计接待游客16.52万人次，同比增长400.13%；旅游收入达826.24万元，同比增长400.17%，获得全国乡村旅游重点村、中国美丽休闲乡村、五椰级乡村旅游点等荣誉称号。

洪水村 图摄 / 孟志军

毛纳村 图摄 / 孟志军

牙胡梯田 图摄 / 孟志军

In the "Tianchuang Community", Maona Village, Shuiman country, Wuzhishan city, the construction of the National Park of Hainan Tropical Rainforest has provided the impetus for accelerating the development of ecological tourism and the Da Ye tea industry. From April 2022 to August 2023, Maonan Village has hosted a cumulative total of 165200 visitors, representing a year-on-year growth of 400.13%. The tourism revenue reached 8.2624 million yuan, a year-on-year growth of 400.17%. The village has been honored with titles such as National Key Village for Rural Tourism, Beautiful and Leisurely Village in China, and Five-Coconut-Level Rural Tourism Site.

探寻雨林乐趣 Explore the Fun of Rainforest

　　海南热带雨林国家公园吊罗山片区正在探索一种特许经营模式，坚持"把该保护的保护好，把该利用的利用好"，在一般控制区适度发展生态旅游、森林康养、林下经济、生态研学等生态产业，提高了一线管护人员的收入，也为国家公园建设可持续发展奠定了坚实基础。

The Douluo Mountain area of the National Park of Hainan Tropical Rainforest is exploring a franchise business model, adhering to the principle of "protect what needs protection and make the best use of what can be utilized". In the general control zone, various ecological industries such as eco-tourism, forest health therapy, understory economic activities, and ecological study programs have been developed in moderation. This has not only increased the income of frontline conservation personnel but has also laid a solid foundation for the sustainable development of the national park.

飞瀑 吊罗山分局供图

"吊罗归来不看水。"

远离尘世纷嚣，倾听飞瀑腾跃，没准儿还能遇见一两种海南特有生物，真是妙不可言。

"Back from Diaoluo , no longer gazing at waters."
Away from the hustle and bustle of the world, listen to sounds of the cascading waterfalls, and you might even encounter one or two of Hainan's unique wildlife species. It's truly indescribably wonderful.

雨林俯瞰 吊罗山分局供图

体验雨林人文
Experience the Rainforest Humanities

2022年11月，在美丽的毛纳村，一场热带雨林与自然、民族、乡村田园交织成章的田园实景剧《雨林时光》正式首演。诗情画意的演艺，将雨林的美、绿色的希望、黎族人的幸福故事娓娓道来，有机会一定要来看看哦！

In November 2022, in the beautiful Maona Village, a picturesque live-action drama titled "Time of the Rainforest" unfolded, weaving together the beauty of the tropical rainforest with nature, culture, and rural life. This artistic performance tells the story of the rainforest's beauty, the hope of green, and the happiness of the Li people in a poetic and narrative manner. If you have the opportunity, you must come and see it!

毛纳村 图摄 / 孟志军

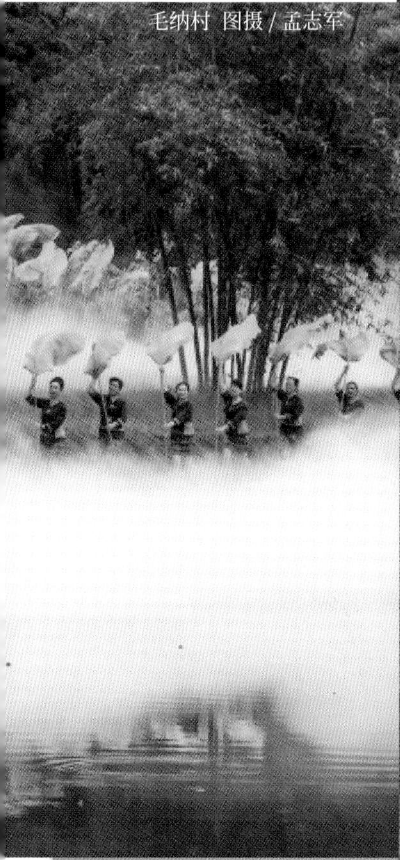
毛纳村 图摄 / 孟志军

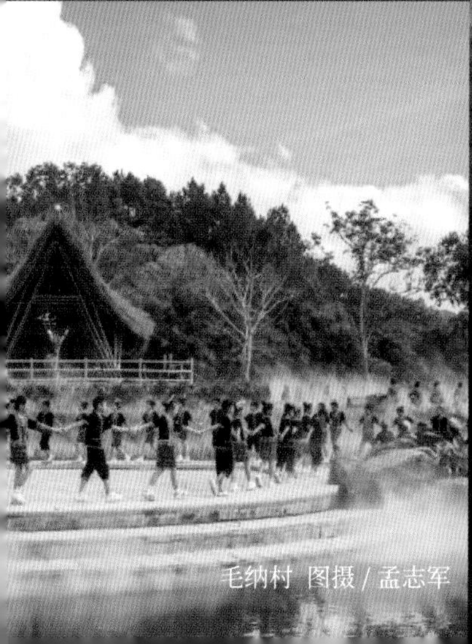
毛纳村 图摄 / 孟志军

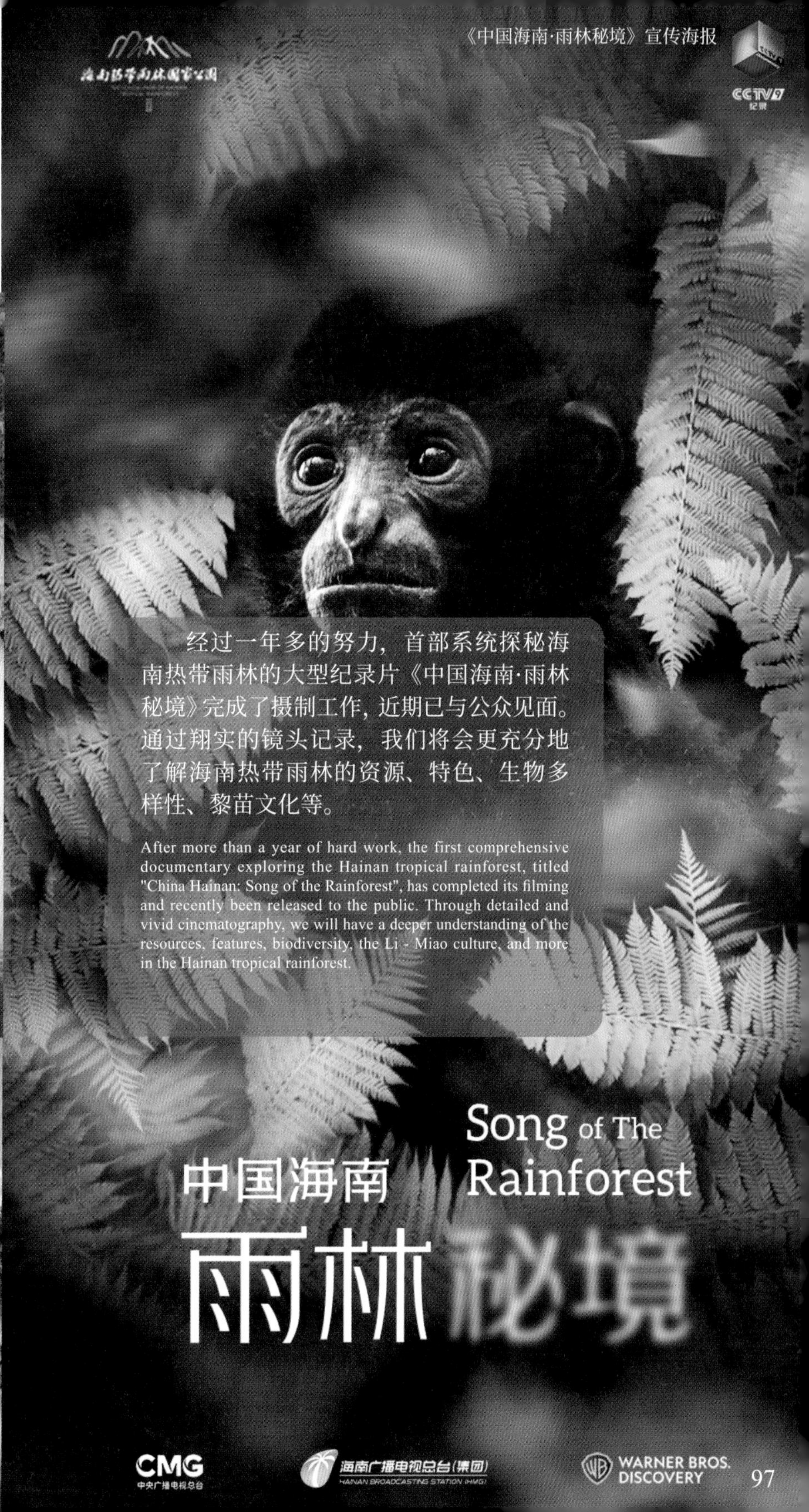

《中国海南·雨林秘境》宣传海报
海南热带雨林国家公园
CCTV9 纪录
经过一年多的努力，首部系统探秘海南热带雨林的大型纪录片《中国海南·雨林秘境》完成了摄制工作，近期已与公众见面。通过翔实的镜头记录，我们将会更充分地了解海南热带雨林的资源、特色、生物多样性、黎苗文化等。
After more than a year of hard work, the first comprehensive documentary exploring the Hainan tropical rainforest, titled "China Hainan: Song of the Rainforest", has completed its filming and recently been released to the public. Through detailed and vivid cinematography, we will have a deeper understanding of the resources, features, biodiversity, the Li - Miao culture, and more in the Hainan tropical rainforest.
中国海南
Song of The Rainforest
雨林秘境
CMG
中央广播电视总台
海南广播电视总台(集团)
HAINAN BROADCASTING STATION (HMG)
WARNER BROS. DISCOVERY

讲好雨林故事 Telling the Rainforest Story

在海南省教育厅、共青团海南省委、海南热带雨林国家公园管理局的共同协商下，统筹协调自然保护、科普教育、生态旅游等功能，无疑成了海南热带雨林国家公园建设的重中之重。

Through joint consultations with the Education Department of Hainan Province, the Hainan Provincial Committee of the Communist Youth League, and the National Park of Hainan Tropical Rainforest Management Bureau, the integrated coordination of functions such as nature conservation, science popularization education, and ecological tourism undoubtedly serves as the core of the construction of the National Park of Hainan Tropical Rainforest.

2020年9月，海南省10所学校挂牌成为第一批海南热带雨林国家公园自然教育学校，目前已策划多场自然科普教育进校园活动，环保志愿者、自然体验导师针对不同年龄段的青少年研发、讲解课程。

In September 2020, ten schools in Hainan Province were designated as the first batch of Hainan Tropical Rainforest National Park Nature Education Schools. They have since organized various environmental education activities within the school campuses. Environmental volunteers and nature experience instructors have developed and delivered courses tailored to young people of different age groups.

自然教育活动 图摄 / 孟志军

自然教育活动 海南国家公园研究院供图

自然教育活动　海南国家公园研究院供图

　　到 2030 年，海南热带雨林国家公园将全面建成天空地一体化等监测监管平台、科技支撑平台、教育体验平台，自然教育受众将达到 300 万人次。热带雨林国家公园将建设成为热带雨林珍贵自然资源传承和生物多样性保护的典范。

By 2030, a comprehensive integrated monitoring and management platform, technological support platform, and educational experience platform will be fully established in the National Park of Hainan Tropical Rainforest. The audience for nature education is expected to reach 3 million. This will transform the National Park of Hainan Tropical Rainforest into an exemplary model for the inheritance of precious natural resources and the preservation of biodiversity in tropical rainforests.

自然教育活动　海南国家公园研究院供图

时隔 24 年，中国再度申报世界"双遗产"

海南"双遗产"申报 海南省林业局供图

2023 年 6 月 10 日，海南省已正式启动"海南热带雨林和黎族传统聚落"世界文化与自然双重遗产（简称"海南'双遗产'"）申报工作。这是自 1999 年武夷山被评为世界文化与自然双重遗产以来，时隔 24 年，中国再度申报。

On June 10, 2023, Hainan Province officially initiated the application for the "Hainan Tropical Rainforest and Li Ethnic Traditional Settlements" as a World Cultural and Natural Double Heritage (referred to as "Hainan Double Heritage"). Since Mount Wuyi was rated as a World Cultural and Natural Double Heritage in 1999, China has applied again after 24 years.

世界遗产是指联合国教科文组织和世界遗产委员会依据《保护世界文化和自然遗产公约》确认的人类罕见的、无法替代的财富，是全人类公认的具有突出意义和普遍价值的文物古迹及自然景观。

World Heritage refers to the human treasures of exceptional and universal value, recognized and confirmed by the United Nations Educational, Scientific and Cultural Organization (UNESCO) and the World Heritage Committee in accordance with the *Convention Concerning the Protection of the World Cultural and Natural Heritage*. These treasures include rare and irreplaceable culture relics and natural landscapes, acknowledged by all of humanity for their outstanding significance and universal value.

After a 24-Year Gap, China Reapplies for World "Double Heritage"

海南热带雨林国家公园周边社区是黎族、苗族群众的聚居地，民族民俗文化丰富多元。2022 年，"海南热带雨林和黎族传统聚落"被联合国教科文组织列入世界遗产预备名单，目前正在紧锣密鼓地推进申报世界文化与自然双重遗产工作。

The communities surrounding the National Park of Hainan Tropical Rainforest are inhabited by the Li and Miao ethnic groups and possess rich and diverse ethnic and folk cultures. In 2022, the "Hainan Tropical Rainforest and Li Ethnic Traditional Settlements" was placed on UNESCO's World Heritage tentative list and is currently making significant progress in the application process for recognition as a World Double Heritage.

黎族文化 图摄 / 孟志军

截至 2023 年，全球范围内有 39 项世界文化与自然双重遗产，中国占 4 项，分别是泰山、黄山、峨眉山 - 乐山大佛、武夷山。泰山是中国也是世界上第一个文化与自然双重遗产，于 1987 年被联合国教科文组织世界遗产委员会列入《世界遗产名录》。

As of 2023, there are a total of 39 World Cultural and Natural Double Heritage sites globally, with China accounting for 4 of them, including Mount Tai, Mount Huang, Mount Emei-Leshan Giant Buddha, and Mount Wuyi. Mount Tai was the first in both China and the world to be recognized as a Natural and Cultural Double Heritage site and was included in the UNESCO World Heritage Committee's *World Heritage List* in 1987.

牙胡梯田 图摄 / 孟志军

世界遗产包括世界文化遗产（包含文化景观）、世界自然遗产、世界文化与自然双重遗产三类。

World Heritage is categorized into three main types: World Cultural Heritage (encompassing cultural landscapes), World Natural Heritage, and Mixed Heritage.

万物有灵，万物共生。

在这片热带雨林的奇妙之境，我们感受到了大自然的变幻莫测，认识了奇妙多彩的动植物，见证了人与自然和谐相处的生态保护之路，也在践行着把绿水青山变成"金山银山"的生态高质量发展之路。

海南热带雨林国家公园是自然的宝库、生命的摇篮。愿每一位读者在这次科普之旅后，都能更加喜欢这片生机盎然的雨林，珍惜大自然的馈赠，努力为生态环境的保护和可持续发展贡献一份力量。

感谢您的阅读，愿自然之美在我们心中长存。

All things possess spirit, and all things coexist.

In this enchanting realm of the tropical rainforest, we have marvelled at the artistry of nature, encountered the wondrous and vibrant flora and fauna, and witnessed the ecological path toward harmonious coexistence between humans and nature. Here, we are actively participating in the journey of transforming lucid waters and lush mountains into invaluable assets, embracing a path of ecologically high-quality development.

The National Park of Hainan Tropical Rainforest is the treasure trove of nature and the cradle of life. May every reader cultivate a deeper appreciation for this thriving rainforest. May each reader cherish the gifts of nature and contribute earnestly to the protection and sustainable development of the ecological environment.

Thank you for joining us on this enlightening journey. May the beauty of nature linger in our hearts.

海南热带雨林 图摄/孟志军